Fundamentals of Energy Efficient Lighting and Controls

RIVER PUBLISHERS SERIES IN ENERGY MANAGEMENT

Series Editor:

MICHELE ALBANO
Aalborg University, Denmark

The "River Publishers Series in Energy Management" is a series of comprehensive academic and professional books focussing on management theory and applications for energy related industries and facilities. Books published in the series serve to provide discussion and exchange information on management strategies, techniques, methodologies and applications, with a focus on the energy industry.

Topics include management systems, handbooks for facility management, safety, security, industrial strategies, maintenance and financing, impacting organizational communications, processes and work practices. Content is also featured for energy resilient and high-performance buildings.

The main aim of this series is to serve as a useful reference for academics, researchers, managers, engineers, and other professionals in related matters with energy management practices.

Books published in the series include research monographs, edited volumes, handbooks and textbooks. The books provide professionals, researchers, educators, and advanced students in the field with an invaluable insight into the latest research and developments.

Topics covered in the series include, but are not limited to:

- Facility management;
- Safety and security;
- Management systems and solutions;
- Industrial energy strategies;
- Financing and costs;
- Energy resilient buildings;
- Green buildings management.

For a list of other books in this series, visit www.riverpublishers.com

Fundamentals of Energy Efficient Lighting and Controls

Jason Livingston LC, IES, IALD
USA

River Publishers

Routledge
Taylor & Francis Group
NEW YORK AND LONDON

Published 2025 by River Publishers
River Publishers
Alsbjergvej 10, 9260 Gistrup, Denmark
www.riverpublishers.com

Distributed exclusively by Routledge
605 Third Avenue, New York, NY 10017, USA
4 Park Square, Milton Park, Abingdon, Oxon OX14 4RN

Fundamentals of Energy Efficient Lighting and Controls / by Jason Livingston LC, IES, IALD.

Routledge is an imprint of the Taylor & Francis Group, an informa business

ISBN 978-87-7004-236-9 (hardback)
ISBN 978-87-7004-806-4 (paperback)
ISBN 978-87-7004-801-9 (online)
ISBN 978-87-7004-800-2 (master ebook)

Dedication

For Jill and Jay

Contents

Preface xvii

About Certified Lighting Efficiency Professional – CLEP™ xix

List of Figures xxi

List of Tables xxv

List of Abbreviations xxvii

1 Lighting Design 1
1.1 Introduction 1
1.2 The Difference Between Lighting Application and Lighting Design 2
1.3 Quality Lighting 4
1.3.1 Human needs 4
1.3.2 Economics and environment 5
1.3.3 Architecture 6
1.3.4 Benefits of quality lighting 6
1.4 The Lighting Design Process 7
1.4.1 Gather project information 8
1.4.1.1 Questions for the owner 8
1.4.1.2 Questions for the architect and interior designer 8
1.4.2 Analyze the lighting requirements and determine lighting goals 8
1.4.3 Select and confirm lighting equipment 9
1.4.3.1 Select luminaires 9
1.4.3.2 Select controls 9
1.4.4 Execute the design 9
1.4.4.1 Lighting design 10

1.4.4.2 Lighting controls 10
1.4.4.3 Construction administration 10
1.5 Designing with Layers of Light 10
1.6 Approaches to Lighting Design Process 13
1.6.1 Pragmatic designs 14
1.6.2 Aesthetic designs 15
1.6.3 Conceptual designs 16

2 Light and Vision **21**
2.1 Introduction 21
2.2 The Electromagnetic Spectrum 22
2.3 Reflection, Refraction, and Diffusion 22
2.3.1 Reflection 23
2.3.1.1 Specular 23
2.3.1.2 Semi-specular 23
2.3.1.3 Matte 24
2.3.2 Refraction 25
2.3.3 Diffusion 25
2.4 Human Vision 26
2.5 Factors Affecting Visual Performance 27
2.5.1 Size 27
2.5.2 Duration 27
2.5.3 Intensity 28
2.5.4 Contrast 29
2.5.5 Age 30
2.6 Scotopic, Mesopic, and Photopic Vision 30

3 Light and Color **33**
3.1 Introduction 33
3.2 Spectral Power Distribution 33
3.3 Chromaticity Diagrams 34
3.4 Color Metrics 37
3.4.1 Color temperature 38
3.4.1.1 Correlated color temperature 39
3.4.2 Duv 40
3.4.2.1 Duv and efficacy 41
3.4.3 LED correlated color temperature 42
3.4.4 LED binning 43
3.4.5 MacAdam ellipses 43

3.5 Color Rendering – CRI 45
3.5.1 Beyond fidelity 48
3.5.1.1 Color shifts 48
3.5.2 Color rendering – ANSI/IES TM-30 50
3.5.2.1 TM-30 measures 52
3.5.2.2 TM-30 design intents and priority levels . 53
3.5.2.3 Designing with TM-30 56

4 Traditional Light Sources 59
4.1 Introduction 59
4.2 Lamp Efficacy and Lamp Life 60
4.3 Incandescent Lamps 60
4.3.1 Standard incandescent lamps 61
4.3.2 Halogen lamps 62
4.3.3 Incandescent and halogen lamp shapes and sizes . . 63
4.3.3.1 Omnidirectional lamp shapes 63
4.3.3.2 Directional lamp shapes 64
4.3.3.2.1 Beam and field angles 66
4.3.3.3 Describing lamp shapes and sizes 67
4.3.3.4 Incandescent lamp bases and sockets . . . 68
4.4 Fluorescent Lamps 68
4.4.1 Linear fluorescent lamps 70
4.4.2 Compact fluorescent lamps 71
4.4.3 Ballasts 72
4.5 High-intensity discharge 72
4.5.1 Metal halide 73
4.5.2 High-pressure sodium 73
4.5.3 Low-pressure sodium 74
4.5.4 HID lamp shapes 74
4.6 Induction 74

5 LED Light Sources 77
5.1 Introduction 77
5.2 Basics of LED Construction and Operation 78
5.3 LED Packages 80
5.3.1 Surface-mounted device 80
5.3.2 Chip-on-board 80
5.4 Organic Light-emitting Diodes 80
5.5 Binning 81

5.6 Drivers . . . 81
5.7 LED Testing and Standards . . . 82
5.8 Environmental Factors . . . 83

6 Luminaires or Lighting Fixtures 85
6.1 Introduction . . . 85
6.2 Classifying Luminaires . . . 85
6.3 Components of New Luminaires . . . 86
6.3.1 Housing . . . 87
6.3.2 Reflectors, flanged or flangeless . . . 87
6.3.3 Beam control – lenses and diffusers . . . 88
6.3.4 Glare control – baffles and louvers . . . 88
6.4 Luminaire Types . . . 88
6.4.1 Downlight . . . 88
6.4.2 Accent or adjustable luminaire . . . 89
6.4.3 Wall washer . . . 90
6.4.3.1 Laying out wall washers . . . 91
6.4.4 Wall grazer . . . 91
6.4.5 Cove light . . . 92
6.4.6 Track and track heads . . . 93
6.4.6.1 Track . . . 93
6.4.6.2 Track heads . . . 94
6.4.6.3 LED tape . . . 95
6.5 Retrofit Lamps and Kits . . . 95
6.6 Luminaire Installation Descriptions . . . 96
6.6.1 Recessed . . . 96
6.6.2 Semi-recessed . . . 97
6.6.3 Surface mount . . . 97
6.6.4 Pendant . . . 97
6.6.5 Poles and bollards . . . 97
6.7 Luminaire Classifications . . . 97
6.7.1 CIE distribution categories . . . 97
6.7.1.1 Direct . . . 98
6.7.1.2 Semi-direct . . . 98
6.7.1.3 Direct-indirect . . . 98
6.7.1.4 Semi-indirect . . . 98
6.7.1.5 Indirect . . . 99
6.7.1.6 Diffuse . . . 99
6.7.2 Outdoor luminaire classifications . . . 99

6.7.2.1 IES cutoff classification 99
6.7.2.2 IES luminaire classification system and BUG ratings 99
6.7.2.3 IES distribution classifications 100
6.7.2.4 NEMA beam angle classifications 101

7 Daylighting 103
7.1 Introduction 103
7.2 Benefits of using Daylighting 104
7.2.1 Reduced energy costs 104
7.2.2 Increases in occupant satisfaction 104
7.2.3 Increased retail sales 105
7.2.4 Enhanced employee well-being and productivity . . 105
7.2.5 Increased student performance 105
7.3 Challenges of using Daylighting 105
7.3.1 Solar heat gain 105
7.3.2 Building heat loss 106
7.3.3 Glare 106
7.3.4 Excessive brightness 106
7.4 The Role of Architects and Lighting Practitioners in Daylighting 106
7.5 Daylighting Techniques 107
7.5.1 Side lighting 108
7.5.2 Top lighting 109
7.6 Daylighting in Energy Conservation Codes 110
7.7 Daylighting Design and Calculation Software Programs . . . 112
7.7.1 Light loss factors in daylighting calculations 113

8 Light Quantity and Calculations 115
8.1 Introduction 115
8.2 Units of Measurement 116
8.2.1 Luminance 116
8.2.2 Illuminance 116
8.2.3 Reflectance 118
8.3 Illuminance Recommendations 118
8.3.1 Modifying illuminance recommendations 119
8.3.1.1 Age of occupants 119
8.3.1.2 Owner's or user's desires 119
8.3.1.3 Design approach 119

8.3.1.4 Alternative recommendations 119
8.3.1.5 Other methods of improving visual performance 120
8.4 Illuminance Calculations 120
8.4.1 Point method 120
8.4.1.1 Accent lighting 125
8.4.2 Lumen method 126
8.4.2.1 Coefficient of utilization 127
8.4.2.2 Light loss factors 130
8.4.2.2.1 Recoverable light loss factors . . 130
8.4.2.2.2 Non-recoverable light loss factors 132
8.4.3 Illuminance calculation software 135
8.5 Spacing Criterion 135
8.6 Photometric Reports 136
8.7 Photometric Files 136
8.8 Photometric Tests 137
8.9 Measurement Standards 137

9 Electricity and Lighting Controls **141**
9.1 Introduction 141
9.2 Electricity Basics 142
9.3 Importance of Controls 143
9.4 Benefits of Controls 144
9.5 Power Controls 144
9.5.1 Switches 145
9.5.2 Relays 145
9.5.3 Dimmers 146
9.5.3.1 Forward phase dimming 147
9.5.3.2 Reverse phase dimming 147
9.5.3.3 Dimming curves 148
9.6 Occupancy and Vacancy Sensors 149
9.6.1 Occupancy vs. vacancy 150
9.6.2 Occupancy detection technologies 151
9.6.2.1 Passive infrared (PIR) detectors 151
9.6.2.2 Ultrasonic detectors 151
9.6.2.3 Microphonic detectors 151
9.6.2.4 Microwave detectors 152
9.7 Photocells 152
9.8 Control Zones and Presets or Scenes 152

9.9 Control Protocols 154
9.9.1 Three-wire control 155
9.9.2 0–10V control 155
9.9.3 DMX and RDM control 156
9.9.4 Digital addressable lighting interface (DALI) 157
9.9.5 Power over Ethernet (PoE) control 157
9.9.6 Wireless control systems 158
9.9.6.1 Wireless devices 158
9.9.6.2 Infrared (IR) communication 159
9.9.6.3 Radio frequency (RF) communication . . 159
9.9.6.4 Bluetooth ® 159
9.9.6.5 Wireless networks 159
9.10 Control Techniques 160
9.10.1 Time scheduling 160
9.10.2 Daylight harvesting 161
9.10.3 Task tuning 161
9.10.4 White tuning 161
9.10.5 Luminaire-level lighting controls 162
9.10.6 Multiple control techniques 162
9.11 Communicating with Other Building Systems 162
9.12 Lighting Controls in Energy Conservation Codes 163
9.13 Documenting the Control System 165
9.13.1 Lighting control schedule 165
9.13.2 Lighting control narrative 165
9.13.3 Lighting control operations schedule 166
9.14 Commissioning 166

10 Light and Health **171**
10.1 Introduction 171
10.2 The Circadian System 172
10.2.1 Circadian entrainment 172
10.2.2 Factors affecting circadian entrainment 173
10.2.3 Measuring circadian lighting effectiveness 174
10.2.3.1 Circadian stimulus 174
10.2.3.2 Equivalent melanopic lux 175
10.2.3.3 Melanopic equivalent daylight illuminance 175
10.3 Ultraviolet Light Disinfection 175
10.3.1 UV disinfection techniques 176
10.3.2 UV dangers and safety 177

10.4 Flicker 178
10.5 Glare 179
10.5.1 Direct glare 179
10.5.2 Reflected glare 180
10.5.3 Disability glare 180
10.5.4 Unified Glare Rating (UGR) 180
10.6 Programs Supporting Occupant Health 181
10.7 Limiting Hazardous Materials 182
10.7.1 Restriction of Hazardous Substances (RoHS) 182
10.7.2 Red List 183

11 Lighting and Sustainability **187**
11.1 Introduction 187
11.2 New Building Sustainability Programs 188
11.2.1 LEED 188
11.2.2 Green Globes 190
11.2.3 BREEAM 190
11.2.4 Sustainability programs for existing buildings 192
11.2.4.1 ENERGY STAR 192
11.2.4.2 CRREM 192
11.3 Product Sustainability 193
11.3.1 Take-back programs 193
11.3.2 Sustainability labeling 193
11.3.2.1 Ingredient health disclosures 193
11.3.2.2 Environmental product declarations 194
11.3.2.3 Declare 194
11.3.2.4 UL 2809-1 environmental claim validation procedure 194
11.4 Qualified Product Lists 196
11.4.1 DesignLights Consortium 196
11.4.2 ENERGY STAR 197
11.5 Other Sustainability Programs 198

12 Lighting System Financial Analysis and Calculations **201**
12.1 Introduction 201
12.2 Initial Cost 202
12.3 Annual Operating costs 202
12.4 Simple Payback 203
12.5 Rate of Return or Return on Investment 205

12.6 Life Cycle Cost Analysis 205
12.7 Utility Bills 207
12.7.1 Administrative, regulatory, and tax charges 207
12.7.2 Delivery charges 208
12.7.3 Supply charges 208
12.7.4 Calculating utility bills 208
12.8 Energy Legislation 208
12.9 Utility Rebates and Government Incentives 209
12.10 Additional Value Considerations 210
12.10.1 Warranty 210
12.10.2 Lighting quality 210
12.10.3 Environmental costs 211
12.10.4 Green programs 211

13 Lighting Maintenance and Environmental Safety **213**
13.1 Introduction 213
13.2 Cleaning and Relamping Basics 214
13.3 Lamp Replacement by Lamp Type 215
13.3.1 Incandescent lamps 215
13.3.2 Halogen lamps 215
13.3.3 Linear and compact fluorescent lamps 216
13.3.4 High-intensity discharge lamps 216
13.3.5 LED lamps 217
13.4 Relamping Strategies 217
13.4.1 Spot relamping 217
13.4.2 Group relamping 218
13.4.3 LED relamping and replacement 219
13.5 Troubleshooting 219
13.5.1 Recurring failures 220
13.5.2 LED retrofit lamps 221
13.6 Environmental Concerns 221
13.6.1 Luminaire components 221
13.6.2 Mercury-containing lamps 221
13.6.3 PCB-containing ballasts 222
13.7 Class 1 and Class 2 Wiring 223

14 Lighting Audits **225**
14.1 Introduction 225
14.2 Lighting Audit Reasons and Results 225

14.3 Audit Types or Levels 226
14.3.1 Level 0 audit 226
14.3.2 Level 1 audit 227
14.3.3 Level 2 audit 227
14.3.4 Level 3 audit 227
14.4 The Lighting Audit Process 227
14.4.1 Level 0 lighting audit 228
14.4.2 Level 1 lighting audit 229
14.4.3 Level 2 lighting audit 230
14.4.4 Level 3 lighting audit 230
14.5 Reports 232

Glossary **235**

Index **257**

About the Author **263**

Preface

This book began as a reference source for those preparing to take the Association of Energy Engineers' (AEE) Certified Lighting Efficiency Professional (CLEP) exam. Early conversations about the contents of this book included an interesting observation: there are several good books aimed at educating future lighting designers (including my own Designing with Light: The Art, Science, and Practice of Architectural Lighting Design) and several good books aimed at educating energy efficiency professionals, but none that address energy efficiency in lighting with an eye toward maintaining quality lighting design. That is the goal of this book.

This organization of this book is similar to other lighting design textbooks, although it takes a more technical, less artistic approach. It was written to:

1) Introduce lighting fundamentals to lighting and energy efficiency professionals
2) Serve as a lighting reference manual for energy professionals engaged in the practice of lighting and
3) Support material taught in the AEE CLEP preparation course for candidates that wish to take the CLEP certification exam.

Throughout this book I speak of "lighting practitioners". There are many people who might design or provide the lighting in a space. The term lighting practitioners includes not just lighting designers but architects, interior designers, electrical engineers, sales representatives, luminaire manufacturers, ESCOs (energy service companies), showrooms, and others who provide some level of lighting design services. This book is addressed to anyone who is interested in learning how to create energy efficient, environmentally sensitive designs that serve the lighting needs of the space's occupants.

Please note that while this book is intended for a global audience I practice in North America. As such, I rely heavily on the authoritative documents produced by organizations such as the Illuminating Engineering Society (IES) and the US Department of Energy, and they are referenced throughout this book. While rarely referenced there are, of course, many other respected authorities around the world who produce a wide range of guidelines, testing procedures, data formats, etc. Readers may need to cross reference these with applicable documents in their country or area.

About Certified Lighting Efficiency Professional – CLEP^{TM}

The Association of Energy Engineer's Certified Lighting Efficiency Professional (CLEP^{TM}) program is designed to provide recognition for professionals who have distinguished themselves as leaders in the field of lighting efficiency. The designation CLEP identifies individuals who have demonstrated high levels of experience, competence, proficiency, and ethical fitness, bringing to their professional activities the full scope of knowledge requisite to the development and implementation of effective lighting efficiency solutions within the commercial, industrial, institutional, and governmental market sectors.

Lighting efficiency professionals understand a building's lighting requirements based on the occupancy and use of a given space, whether the application is illuminating a new building or retrofitting existing buildings to the most current lighting technologies. They measure, verify, and evaluate lighting efficiency projects, including the human-factor implications, technologies, and financial incentives to calculate the return on investment and compare savings against other energy efficiency projects.

Recently the Association of Energy Engineers decided to expand its offering of the Certified Lighting Efficiency Professional beyond the Unites States and simultaneously update the material in the exam preparation course to include LEDs and other new technologies. Initially, an AEE CLEP committee developed a Body of Knowledge course content outline. Next, three CLEPs from across the world were invited to produce the necessary educational materials. This resulted in two versions of the course, for those in the US and those outside the US. The three lighting and energy professionals who developed the materials for the CLEP exam prep course and serve as course instructors are Alain Aoun from Lebanon (and now Canada), Louis Lagrange from South Africa, and Bob Henderson from the United States.

List of Figures

Figure 1.1	Elements of quality lighting.	5
Figure 1.2	Layers of light.	11
Figure 1.3	Examples of focal glow in retail.	12
Figure 1.4	Examples of ambient luminescence in an office.	12
Figure 1.5	Example of play of brilliants.	13
Figure 1.6	Office with a pragmatic lighting design.	15
Figure 1.7	Restaurant with an aesthetic design.	16
Figure 1.8	Example of a meandering river.	17
Figure 1.9	Final installation of meandering river inspired conceptual design.	18
Figure 2.1	Electromagnetic spectrum.	22
Figure 2.2	Specular reflection.	23
Figure 2.3	Semi-specular reflection.	24
Figure 2.4	Matte reflection.	24
Figure 2.5	Refraction.	25
Figure 2.6	Diffusion.	26
Figure 2.7	Parts of the human eye.	26
Figure 2.8	Measured intensity vs. perceived brightness.	28
Figure 3.1	Spectral power distribution for an incandescent light source.	34
Figure 3.2	CIE 1931 (x, y) chromaticity diagram with blackbody locus.	35
Figure 3.3	CIE 1960 (u, v) chromaticity diagram with blackbody locus.	36
Figure 3.4	CIE 1976 (u', v') chromaticity diagram with blackbody locus.	37
Figure 3.5	SPDs of a blackbody radiator at three color temperatures.	38
Figure 3.6	SPD of the blackbody radiator and other light sources. (a) Fluorescent lamp at 3000 K (b) metal halide lamp at 3000 K (c) LED lamp at 3000 K.	39

Figure 3.7	$V(\lambda)$.	41
Figure 3.8	ANSI/NEMA C78.377 CCT quadrangles and center points in the CIE 1976 (u', v') chromaticity diagram.	42
Figure 3.9	MacAdam ellipses (shown at 10× size).	44
Figure 3.10	The eight colors used to calculate CRI R_a (top) and six supplemental colors, including R_9 (bottom).	46
Figure 3.11	Hue and chroma shifts.	49
Figure 3.12	SPDs producing hue and chroma shifts.	49
Figure 3.13	TM-30 simple report.	51
Figure 3.14	A pizza illuminated with (a) a high-fidelity light source (b) a high-preference light source (c) a high-vividness light source.	54
Figure 4.1	Parts of an incandescent lamp.	62
Figure 4.2	Halogen capsule in an A-shaped envelope.	63
Figure 4.3	Omnidirectional lamp shapes. (a) A-lamp (b) B-lamp (c) C-lamp (d) CA-lamp (e) G-lamp (f and g) T-lamps (not to scale).	64
Figure 4.4	Directional lamp shapes. (a) R lamp (b) PAR lamp (c) MR lamp (d) AR lamp (not to scale).	65
Figure 4.5	Beam and field angles.	67
Figure 4.6	Incandescent and halogen lamp bases (not to scale).	69
Figure 4.7	Linear fluorescent lamp shapes. (a) linear (b) u-bend (c) circle-line (not to scale).	70
Figure 4.8	Linear fluorescent lamp bases (not to scale).	70
Figure 4.9	(a) Single U (PL or Biax) (b) double U (PL-C or Double Biax) (c) triple U (PL-T or Triple Biax) (d) twist or swirl retrofit (not to scale).	71
Figure 4.10	Compact fluorescent lamp bases (not to scale).	71
Figure 4.11	Parts of a metal halide lamp.	73
Figure 4.12	HID lamp shapes (not to scale).	74
Figure 4.13	HID lamp bases (not to scale).	75
Figure 5.1	Schematic diagram of a white light LED.	79
Figure 5.2	SPD of a pc-LED showing the characteristic spike of energy from the blue LED and a broad distribution of wavelengths from the phosphor.	79
Figure 6.1	Common luminaire components.	86
Figure 6.2	(a) Recessed 2 × 2 downlight. (b) Surface mount cylinder downlight. (c) Recessed multi-cell downlight (not to scale).	89

Figure 6.3	(a) Internal components of accent or adjustable luminaire. (b) Recessed adjustable luminaire from below. (c) Recessed adjustable multiple (not to scale).	89
Figure 6.4	Section through a lensed wall wash luminaire.	90
Figure 6.5	Wall washer spacing.	91
Figure 6.6	Section through a wall graze luminaire.	92
Figure 6.7	Cove luminaires (a) in a rectangular extrusion and (b) in a knife-edge extrusion (not to scale).	93
Figure 6.8	(a) Single-circuit track. (b) Two-circuit track (not to scale).	94
Figure 6.9	(a) Track head. (b) Track head with cross baffle (not to scale).	94
Figure 6.10	LED tape.	95
Figure 6.11	Extrusions for LED tape.	95
Figure 6.12	CIE distribution categories.	98
Figure 6.13	BUG zones.	100
Figure 6.14	IES outdoor illumination distribution patterns.	101
Figure 7.1	Daylight distribution through a window.	108
Figure 7.2	Daylight distribution using a light shelf.	109
Figure 7.3	Skylight distribution. (a) Straight-sided skylight. (b) Slope-sided skylight. (c) Skylight with a bottom fiffuser.	110
Figure 7.4	Daylighting side-lit zone adjacent to a window. (a) Plan view. (b) Section view.	111
Figure 7.5	Daylighting zone under a rooftop monitor. (a) Plan view. (b) Section view.	111
Figure 7.6	Daylighting zone under a skylight. (a) Plan view. (b) Section view.	112
Figure 8.1	Footcandle and lux.	117
Figure 8.2	Inverse square law.	121
Figure 8.3	Candelas table and candlepower distribution graph.	123
Figure 8.4	Measuring angle of incidence.	124
Figure 8.5	Example coefficient of utilization table.	128
Figure 8.6	Room cavities.	129
Figure 9.1	Alternating and direct currents.	142
Figure 9.2	(a) Paddle or "Decora" switch. (b) Toggle switch with sliding dimmer. (c) Rotary dimmer. Courtesy of Lutron Electronics Company, Inc.	145

Figure 9.3 (a) 4-relay module. (b) 48-relay panel with circuit breakers (not to scale). Courtesy ETC, Inc. 146
Figure 9.4 Forward Phase Dimming. 147
Figure 9.5 Reverse phase (ELV) dimming. 148
Figure 9.6 Logarithmic dimming output produces linear change in brightness. 149
Figure 9.7 (a) Wall-mounted occupancy sensor. (b) Dimmer switch with built-in occupancy sensor. (c) Ceiling-mounted occupancy sensor (not to scale). 150
Figure 9.8 A conference room with at least six zones of lighting. 153
Figure 9.9 (a) Preset control station. (b) Preset control station with dimmer controls (not to scale). Courtesy of Lutron Electronics Company, Inc. 154
Figure 9.10 Touchscreen controller. Courtesy of ETC, Inc. . . . 154
Figure 10.1 Upper room air disinfection with a wall-mounted UV luminaire. 177
Figure 11.1 Example of a Declare label. 195
Figure 14.1 Occupancy/lighting logger. 228

List of Tables

Table 3.1 Nominal LED correlated color temperatures used in general lighting. 40
Table 3.2 Summary of TM-30 measures (Royer, 2022). 52
Table 3.3 TM-30 Annex E design intent and priority level descriptions. 55
Table 4.1 Light sources, efficacy, and average life. 61
Table 4.2 Beam angle descriptions and designations. 66
Table 6.1 IES cutoff classifications. 99
Table 6.2 NEMA beam spread classifications. 101
Table 8.1 Accent illuminance ratios. 126
Table 8.2 Typical lamp lumen depreciation values. 130
Table 9.1 Sample lighting control schedule. 166
Table 9.2 Sample lighting control operations schedule. 166
Table 12.1 Simple payback worksheet. 204
Table 12.2 Simple payback comparison of three lighting systems. 204
Table 12.3 Simple life cycle cost analysis worksheet. 206
Table 12.4 Utility bill calculation. 208
Table 14.1 Level 1 audit form. 229
Table 14.2 Level 2 audit form. 231

List of Abbreviations

AC	Alternating current
ACGIH	American Conference of Governmental Industrial Hygienists
BF	Ballast factor
BIM	Building information modelling
BMS	Building management system
BREEAM	Building Research Establishment Environmental Assessment Methodology
CBCP	Center beam candlepower
CCT	Correlated color temperature
cd	Candela
CES	Color evaluation sample
CFL	Compact fluorescent
CIE	Commission Internationale de lâĂŹEclairage
CMH	Ceramic metal halide
COB	Chip-on-board
cp	Candlepower
CRI	Color rendering index
CRREM	Carbon Risk Real Estate Monitor
CS	Circadian stimulous
CT	Color temperature
CU	Coefficient of utilization
CVG	Color vector graphic
DALI	Digital addressable lighting interface
DC	Direct current
DGNB	German Sustainable Building Council
DLC	DesignLights Consortium
DMX	Digital multiplex for lighting
ELV	Electronic low voltage
EML	Equivalent melanopic lux
EPD	Environmental product declaration
ESG	Environmental, social, and corporate governance
fc	Footcandle

HID	High-intensity discharge
HIR	Halogen infrared
HPS	High-pressure sodium
HVAC	Heating, Ventilation and Air Conditioning
IES	Illuminating Engineering Society
ILV	International Lighting Vocabulary
ipRGC	Intrinsically photosensitive retinal ganglion cell
kW	Kilowatt
kWh	Kilowatt hour
LBO	Lamp burnout
LCA	Life cycle assessment
LDD	Luminaire dirt depreciation
LED	Light-emitting diode
LEED	Leadership in Energy and Environmental Design
LLD	Lamp lumen depreciation
lm	Lumen
lm/W	Lumens per Watt
LPD	Lumen power density
LPS	Low-pressure sodium
LRC	Lighting Research Center
lx	Lux
MBCP	Middle beam candlepower
M-EDI	Melanopic equivalent daylight illuminance
MH	Metal halide
NEMA	National Electrical ManufacturerâĂŹs Association
OLED	Organic light emitting diode
PAR	Parabolic aluminized reflector
PCB	Polychlorinated biphenyls
PC-LED	Phosphor converted LED
PIR	Passive infrared
PoE	Power over Ethernet
PPFD	Photosynthetic photon flux density
QPL	Qualified product list
RCR	Room cavity ratio
RFI	Request for information
RoHS	Restriction of Hazardous Substances
ROI	Return on investment
RP	Recommended practice

RSDD	Room surface dirt depreciation
SC	Spacing criteria
SCN	Suprachiasmatic nucleus
SCR	Silicon-controlled rectifier
SDCM	Standard deviation color matching
SI	International System of Units
SMD	Surface-mounted device
SPD	Spectral power distribution
TCLP	Toxicity Characteristic Leaching Procedure
TF	Thermal factor
TLV	Threshold limit value
TRIAC	Triode alternating current
UDP	User datagram protocol
UGR	Unified Glare Rating
UV-C	Ultraviolet C
VT	Visible transmission

1

Lighting Design

Abstract

One aspect of lighting design involves weighing the competing demands of issues such as energy consumption, budget, aesthetics, etc. This chapter introduces the balance between energy efficiency and quality lighting design, and the difference between a lighting design and a lighting application. It includes an overview of the lighting design process, and the components and the benefits of quality lighting designs. The chapter concludes with two frameworks used by lighting practitioners: Richard Kelly's Forms of Lightplay and Jason Livingston's Design Approaches.

Keywords: Lighting design, lighting application, focal glow, ambient luminescence, play of brilliants, Richard Kelly, pragmatic design, aesthetic design, conceptual design, elements of quality lighting, lighting design process, layers of light.

1.1 Introduction

The goal of this book is to dispel the idea that energy-efficient lighting design and aesthetically pleasing lighting design are at odds with one another. There should be no separation. In fact they should be considered as one: A quality lighting design is energy efficient, and effectively meets a range of needs, including aesthetic needs, of the owner and users. The aim of this chapter is to help energy efficiency professionals understand the aesthetic and design aspects of lighting so that their work results in liveable environments, not just energy-efficient ones.

The most energy-efficient lighting design is also likely to be the one of the worst lighting designs for occupants of the space, for several reasons. First, the most optically efficient fixtures place the LEDs as close to the fixture aperture as possible (Chapter 6), increasing the likelihood of creating glare (Chapter 8). Second, the most energy-efficient white light LEDs are likely to have a high color temperature and poor color rendering (Chapter 3), which may be unpleasant or even unacceptable to the occupants and/or for the application and may have health impacts on the users (Chapter 10). Third, the design would only provide the light needed for each task (Chapter 8) at each task location, ignoring the balance and volume of the room. Which means, finally, the design would create a utilitarian environment that ignores opportunities to support or enhance the architecture and interior design, or to provide users with an aesthetically pleasant and visually interesting space.

On the other hand, a lighting design that focuses exclusively on aesthetics could be an equally poor design, but for different reasons. It might use inefficient incandescent lamps (Chapter 4) in inefficient fixtures (Chapter 6) and ignore opportunities to take advantage of daylight (Chapter 7). To simplify operations and remove the visual clutter of controls (Chapter 9), it would sacrifice the associated energy savings. It would, therefore, waste the owner's money (Chapter 12) and be difficult and expensive to maintain (Chapter 13).

A high-quality, energy-efficient lighting design must balance the competing interests of using energy-efficient light sources and fixtures, providing appropriate lighting for visual tasks, supporting the architectural and interior designs and the room's aesthetics, creating visual interest in the space, and meeting all applicable construction, accessibility, and energy-efficiency codes – all while staying on budget. In addition, the design may be required to exceed code requirements in specific ways (as when complying with a green program such as Leadership in Energy and Environmental Design (LEED)) or support the health and well-being of the occupants (as when complying with a program such as WELL Building).

1.2 The Difference Between Lighting Application and Lighting Design

A lighting application is the project's category or type, such as office, retail, or healthcare. Several professional organizations and manufacturers publish documents that describe the lighting requirements of various applications and provide guidance on how to meet them. However, meeting an application's requirements does not guarantee a quality lighting design.

Broadly speaking, when faced with the same design challenge designers and engineers will develop different solutions using different criteria. Both will look at their completed project as being successful because it met or exceeded the criteria they developed. Both might look at the work of the other and see failure because their own criteria were not met.

For example, the three primary goals of a retail lighting design are to provide light that attracts and guides the customer, allows the customer to evaluate the merchandise, and facilitates the completion of the sale (IES Retail Lighting Committee, 2020). Each of these goals has a recommended brightness level that can be identified and met, and a lighting design that strictly fulfills these goals might simply be bright – bright enough to attract a customer's attention, bright enough for the customer to see and evaluate the merchandise, bright enough for the customer and salesperson to complete the purchase. This approach – of meeting the minimal requirements of the lighting application – is the one taken by many low-cost, big box retailers. The light provided meets the application's requirements for visibility although the shopping experience may not be aesthetically pleasing. One might think of this as an engineering solution to the lighting application. It fulfills the application's technical requirements with little regard for aesthetics.

Design-focused lighting practitioners of higher-end retail projects, on the other hand, look at the same requirements and consider how using other aspects of light – not just brightness but color temperature and color, distribution, and change or movement of light over time – can be used to simultaneously meet illumination requirements and heighten the shopping experience. For example, lower levels of ambient illumination permit an accent light to draw attention to a featured display, light that has directionality (and therefore creates highlights and shadows that enhance three-dimensionality) reveals the texture of surfaces, higher levels of light at service areas guide shoppers to check-out counters, etc. This approach is likely to require more luminaires (light fixtures) and therefore cost more, and may consume more energy, but the result is a more pleasant experience – one that is less like a warehouse and more like a destination – which may be more beneficial to the owner, employees, and customers than a strict focus on energy efficiency.

This sums up the first challenge of lighting design and hints at the first steps that must be taken in providing a high-quality design. The resulting light must meet the full range of requirements set out by the owner, design team, lighting practitioner, and others, not only application requirements or energy efficiency or aesthetics. The lighting design process needed to meet all of these demands can be summarized as follows:

1. Understand not just the project type and the application (such as light an office) but the various wants and needs of the owner, users, architect,

interior designer, and lighting practitioner. Those wants and needs are likely to cover a wide range of goals including cost, time frame, aesthetics, energy efficiency, controllability, and flexibility.
2. Gather relevant information toward those goals.
3. Reconcile conflicts among those goals. The trade-off for fixtures that are readily available and inexpensive, for example, may be that they are unattractive or have poor beam control and are likely to produce glare.
4. Develop a set of design criteria for success in meeting all of the project's lighting goals or prioritize goals if they can't all be met equally.
5. Develop a design that is guided by the design criteria but that is open and flexible enough to allow for growth, change, and improvement throughout the design process.

In other words, "to play with light is to play with magic – it demands (1) a trained eye to recognize real and relative values (2) experience and knowledge of the cultural and psychological effects of light on people (3) experience and knowledge of physical techniques" (Kelly, 1952).

1.3 Quality Lighting

After more than a century of research, we have a clear understanding of the amount of light needed by the average person for a wide range of visual tasks, from aircraft assembly to playing volleyball. There is also a small, but growing, body of studies that examine other aspects of light and lighting, from circadian entrainment (keeping the body's clock aligned with day/night, see Chapter 10) to the environmental impact of the production, use, and disposal of a luminaire (Chapter 11). The question is how to synthesize this information to identify and describe a quality lighting design. One framework looks at how the design's components fit into three overlapping elements: the users of the space, the economics of the installation, and the installation's relationship to the building, as shown in Figure 1.1 (Veitch, 2018, IES Light and Design Committee, 2020).

While not every item shown will get equal weight, and some may have no relevance in certain types of projects, these three short lists summarize the elements that should be considered in the pursuit of quality lighting.

1.3.1 Human needs

Quality lighting design should meet human needs by:

- providing appropriate illumination for tasks;

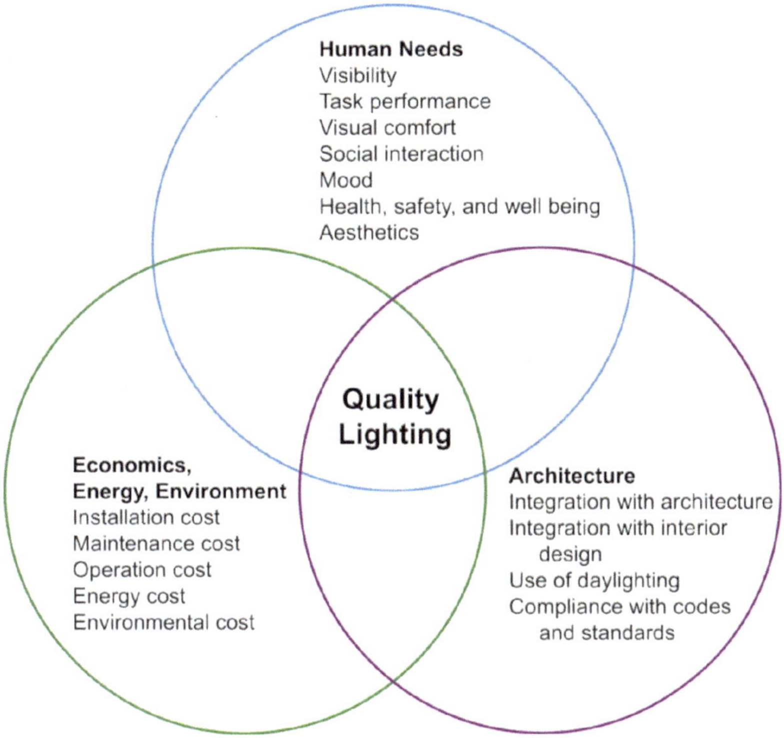

Figure 1.1 Elements of quality lighting.

- providing ambient illumination that delivers light of appropriate uniformity, illuminates vertical as well as horizontal surfaces, and is free of glare;
- properly modeling faces and other important three-dimensional objects in the room;
- using color temperature and color rendering that are appropriately coordinated with the interior finishes, occupant activities, function of the space, and desired mood/atmosphere.

1.3.2 Economics and environment

Quality lighting is conscious of the owner's costs, and of the costs to the environment by:

- being easy to install and service so that the original design intent is maintained;

- providing a low cost of ownership by using energy-efficient luminaires, meeting energy codes, and employing controls to maximize energy savings;
- being environmentally sensitive in conserving natural resources by selecting energy-efficient luminaires with recycled content and/or that are recyclable, and environmentally aware by knowing the origin and contents of luminaire components.

1.3.3 Architecture

Quality lighting is designed with attention to the architecture and interior design so that it:

- is integrated with the architecture – appropriately standing out, blending in, or accenting architectural features;
- is integrated with the interior design by providing color temperature and color rendering that support the interior color and materials palettes;
- provides visual interest, variation, and hierarchy;
- uses daylight wherever appropriate;
- complies with applicable building, electrical, and conservation codes.

1.3.4 Benefits of quality lighting

While those focused on energy efficiency or lowest cost might view an emphasis on other aspects of design as an extravagance, wasteful, or an inappropriate use of energy or other resources, it is important to remember that ultimately spaces are illuminated for people. People need to engage with their surroundings and lighting design can support, and even be a part of, that engagement. A long list of studies has shown that high-quality lighting design has many benefits that offset energy use that is higher than strictly needed to achieve an application's recommended light levels.

Studies have shown that quality lighting can deliver benefits to individuals occupying buildings, the organizations owning and operating buildings, and the environment. Demonstrated benefits include:

Improved student test scores (Heschong, 2013). This study looked at 21,000 students in three distinct school districts. The findings indicate better student test scores in daylit classrooms and suggest that there is an important relationship between daylight availability in buildings and human performance.

Improved student concentration (Sleegers et. al., 2012). This study varied the lighting for elementary school children from two schools. It found that the lighting system can have a positive influence on students' concentration.

Improved sleep–wake cycles for adults with dementia (Figueiro, MG, et. al., 2023). This study demonstrated that light therapy focused on the circadian system by delivering bright days and dark nights improved measures of sleep, including sleep duration, and mood.

Increased retail sales. Multiple studies have found that retail stores with daylight provided by skylights have retail sales 10% (Ellis, 2007) to 40% (Heschong, 2003) higher than similar stores without daylight.

Enhanced employee well-being (Figueiro & Rea, 2016). This study on circadian lighting showed that people working in offices with more daylight exposure self-reported better sleep quantity and quality and a more positive mood.

Enhanced employee well-being and productivity (Veitch et al., 2010). This study showed that direct/indirect lighting with individual control is perceived as being more comfortable than recessed parabolic luminaires. The study also discusses previous research showing that people who judged the lighting more comfortable also reported that the room was more pleasant, and their work performance satisfaction was higher, which predicts higher commitment to their job and lower intent to leave their employer.

Enhanced employee productivity (Veitch, 2018). This study found that providing high-quality lighting enables a productive workplace, and that ensuring employees' well-being both reduces costs and increases output.

1.4 The Lighting Design Process

While lighting practitioners go through the same phases of design and construction as the rest of the design team (programming, schematic design, design development, construction documentation, construction administration) those phase names don't convey the work that happens during each phase. In addition, it is not uncommon for the lighting designer's work to begin after the architect's, in design development for example, so for some time the lighting designer's work will be out of phase with other team members. One framework for the lighting design process, from start to finish, is described below (Livingston, 2022).

1.4.1 Gather project information

Before the design can begin, the designer needs to understand the needs, wants, and goals of the various stakeholders. These questions may not all be answered at the beginning of the project, so they may be addressed during programming, schematic design, and/or design development. Questions for the relevant stakeholders include:

1.4.1.1 Questions for the owner

- What is the purpose of the project?
- What are your general goals? Are there any specific goals?
- Do the occupants of the space (employees, customers, guests) have any known needs or desires?
- What are the visual tasks that are performed by the occupants?
- What are your goals for aesthetics, mood, etc?
- Are you asking the design team to achieve any sustainability or wellness goals, or follow any voluntary programs (such as LEED or WELL Building)?

1.4.1.2 Questions for the architect and interior designer

- What are the various codes that apply to this project?
- What are your goals for aesthetics, mood, etc.?
- Which spaces are you designing for a single activity and which spaces are for multiple activities or are flexible?
- What are the color and materials palettes?

1.4.2 Analyze the lighting requirements and determine lighting goals

Analysis can't happen without relevant information, but once the information is gathered the project's lighting requirements and goals can be determined, including:

- The general lighting requirements for each space.
- The light levels and contrast ratios for the visual tasks performed in each space, and for the entire project.
- The role lighting will play in creating and/or supporting the aesthetics/mood/atmosphere of each space.
- Based on the information above, the appropriate luminaire types and lighting techniques to be used.

- The maximum power consumption permitted by the energy conservation code.
- The lighting controls required by the building codes, and the owner/user desires for controls.
- Lighting control interface with other systems, such as audio-video or building management.

1.4.3 Select and confirm lighting equipment

Once lighting practitioners fully understand what is required, they can turn to the question of how to achieve those goals.

1.4.3.1 Select luminaires

- Determine the distribution pattern for each type of luminaire.
- Determine beam angles and focusing angles.
- Select appropriate luminaires and options.
- Determine mounting details and communicate special details, such as coves and soffits, to the architect or interior designer.
- Create preliminary layouts to confirm the design intent and determine luminaire quantity.
- Confirm luminaire availability and cost.

1.4.3.2 Select controls

- Confirm the control system can interface with other systems as required.
- Confirm the control system can output the required luminaire control protocol(s) (0-10V, DMX, DALI, etc.)
- Select appropriate occupancy/vacancy sensors.
- Select appropriate photosensors.
- Select appropriate user interface(s), such as push buttons or touch-screens.
- Create preliminary layouts to confirm design intent and component quantity.
- Confirm equipment availability and cost.

1.4.4 Execute the design

Once preliminary selections and layouts are determined, the design process becomes one of revisions, additions, and deletions as the design is adapted to changes by other design team members and is refined by the lighting practitioner.

1.4.4.1 Lighting design

- Confirm the luminaires and layout meet light level goals.
- Confirm total luminaire wattage is within the limits set by the building or energy conservation code.
- Create final lighting plans and details.
- Create final luminaire schedule and cut sheets.
- Create final luminaire specifications.

1.4.4.2 Lighting controls

- Confirm interface with other systems.
- Confirm luminaire control protocol(s).
- Create final risers, plans, and details.
- Create final zoning and device schedules, and device cut sheets.
- Create final control narrative.
- Create final control specifications.

1.4.4.3 Construction administration

- Attend project meetings with the contractor.
- Respond to requests for information (RFIs)
- Review shop drawings or submittals for luminaires and controls.
- Make progress site visits and inspections.
- Focus all focusable luminaires.
- Confirm control system commissioning and set light levels in all presets.

1.5 Designing with Layers of Light

A lighting design with a single set of fixtures, recessed downlights for example, ignores opportunities to accent architecture or provide visual variety. Such a design is inevitably monotonous and is likely to be described with words ranging from dull to oppressive. Thoughtful, high-quality lighting designs use overlapping sets of fixtures to provide lighting that fulfills multiple requirements, such as ambient lighting and task lighting, and creates a visually engaging environment. This approach is called layering, and each lighting technique used is a layer of light. For example, Figure 1.2 has at least five layers of light.

The most well-known description of layering comes from the pioneering lighting designer Richard Kelly who described part of his design process of using three layers of light. He referred to these layers as three forms of

Figure 1.2 Layers of light.
iStock.com/poplasen

lightplay, and called them focal glow, ambient luminescence, and play of brilliants (Kelly, 1955).

Focal glow (Figure 1.3) describes pools of light that are brighter than the surrounding area, which draw the attention of people in the space. Kelly describes focal glow as "the campfire of all time, the glowing embers around which stories are told, or the football rally bonfire. The attraction of focal glow commands attention and creates interest. It fixes the gaze, concentrates the mind and tells people what to look at. It sells merchandise. Focal glow separates the important from the unimportant. It establishes precedence, can induce movement, and direct and control traffic." Focal glow is a layer that uses brightness to draw attention.

Ambient luminescence (Figure 1.4) is even, shadowless illumination. "Ambient Luminescence is a snowy morning in the open country. It is... a white tent at high noon. Ambient luminescence minimizes the importance of all things and all people. It fills people with a sense of freedom of space and can suggest infinity. It is usually reassuring and restful." Ambient luminescence, which may be bright or dim, is the layer that evenly blankets a space with light.

Figure 1.3 Examples of focal glow in retail.
iStock.com/zhudifeng

Figure 1.4 Examples of ambient luminescence in an office.
iStock.com/peshkov

Figure 1.5 Example of play of brilliants.
Courtesy of Ed McCarthy

Play of brilliants (Figure 1.5) is the sparkle that comes from decorative light fixtures. "Play of Brilliants is... a cache of diamonds in an open cave or the Versailles hall of mirrors with its thousands of candle flames – a ballroom of crystal chandeliers. Play of brilliants is Times Square at night. It is birch trees interlaced by a car's headlights. Play of brilliants excites the optic nerve, in turn stimulates the body and spirit and charms the senses. It creates a feeling of aliveness, alerts the mind and awakens curiosity, and sharpens the wit." Focal glow is the decorative lighting layer.

Interestingly, all three forms of lightplay refer, primarily, to the distribution of light or how light is spread across a space. Unaddressed, but still of critical importance, are other qualities of light such as brightness, directionality, and color temperature (the warmth or coolness of white light). Nonetheless, the fact that Kelly's approach is still widely used more than 70 years after he articulated it speaks to the importance of using multiple layers in lighting designs in most lighting applications.

1.6 Approaches to Lighting Design Process

A lighting practitioner gathers information about the wants and needs of the different stakeholders on a project, resolves any conflicts among those wants and needs, establishes design criteria, and creates and documents the design. Often, when referring to the criteria and execution, the lighting practitioner refers to their "lighting concept," although a concept is not always required. A design concept is an abstract idea that is used to guide the design toward a particular outcome. Some projects are simpler and can be effectively carried out without a full "concept." There are three approaches that a lighting designer can take – pragmatic, aesthetic, or conceptual (Livingston, 2014).

1.6.1 Pragmatic designs

In a pragmatic approach, the designer identifies the requirements of the project and any potential problems with meeting them, resolves the problems, and executes the design. As each element of the design is completed it is checked against the others (lighting, architectural, and interiors) to be sure that the design is comprehensive and in harmony with the other disciplines.

For example, in a pragmatic approach to lighting an office (Figure 1.6) the lighting practitioner might do the following:

- Identify the recommended light levels for desktops and other task surfaces. If it is appropriate, modify those recommendations to meet the expressed needs or desires of the owner and/or users. Provide a combination of ambient and task lighting to meet those light levels.
- Examine the architecture and interior design for opportunities to highlight interesting features such as alcoves, columns, or textured walls. If none exist, consider creating them by washing walls with light, adding decorative pendants or wall sconces, accenting any art or graphics on the walls, and/or talking to the architect or interior designer about modifying their design by, for example, adding coves to certain ceilings.
- If the owner requests that the design meet energy, green, or wellness goals such as LEED or WELL Building certification, identify the achievable points in the system, communicate them to the design team and the owner, and work toward meeting those requirements.
- Calculate the total load (wattage) that is permitted by code and ensure that the design is under that limit.
- Confirm the code requirements for lighting controls as well as any requirements of the owner and/or users. Select a control system that meets those requirements, and design and document the system.

The resulting design couldn't be described as conceptual or having a concept, but it would be a code compliant, energy efficient, visually interesting

Figure 1.6 Office with a pragmatic lighting design.
iStock.com/monkeybusinessimages

design. For some projects and some clients that's exactly what is desired and required.

1.6.2 Aesthetic designs

Aesthetic designs focus on creating or supporting a mood, ambiance, or environment (Figure 1.7) while also meeting the pragmatic requirements of the project. Aesthetic designs are often guided by a set of adjectives that describe the desired result, such as, "The restaurant should feel like a romantic, mysterious Italianate grotto."

While ensuring that pragmatic requirements are met, a lighting designer/practitioner taking an aesthetic approach to lighting a romantic, mysterious Italianate grotto might do the following:

- Meet with the owner, architect, and interior designer to define what is meant by "romantic, mysterious Italianate grotto." This may include compiling a list of adjectives to describe the project or agreeing on a group of photos that illustrate the desired lighting. For the architect and

Figure 1.7 Restaurant with an aesthetic design.
Photo by Igor Rand on Unsplash

interior designer, it may include a list of materials, colors, and/or textures that should be used.

- Closely examine the architectural and interiors drawings looking for opportunities to integrate decorative fixtures, hidden fixtures, and generally create a holistic design for the restaurant.
- Search manufacturers websites for decorative fixtures with the appropriate style and materials. Obtain samples if necessary. Present the fixtures to the owner and design team for discussion, approval, and integration into the project.
- Perform illuminance calculations to determine if the decorative fixtures and their locations are sufficient, or if additional illumination (perhaps from very discrete downlights) is needed.

The resulting restaurant wouldn't have a lighting concept per se, but would have a very strong, clear mood and ambience that was the result of the design team working closely together toward a common aesthetic goal.

1.6.3 Conceptual designs

Conceptual designs typically use a simile or metaphor, or a visual or mental image, to guide the development of the design. The concept statement presents the desired quality of the final design without being prescriptive in what it is or how it is achieved. For example:

Figure 1.8 Example of a meandering river.
iStock.com/wonganan

- In a high-tech business-to-business sales center the architect suggested using natural materials and biophilic forms as a counterpoint to the futuristic presentation material and the owner agreed.
- After much consideration and research, the lighting designer proposed a series of custom curvilinear light fixtures whose shapes would be based on meandering rivers, as shown in Figure 1.8. Both the owner and architect were excited by the concept and agreed to it.
- The lighting designer further proposed that the meandering fixture would use red/green/blue/white LEDs so the fixture could produce white light, but also colored light for events or parties that where hosted in the space. This was also agreed to by the owner and architect.
- The lighting designer prepared and tested multiple layouts to cover the space, provide appropriate light levels, and avoid conflict with air ducts and other building systems.
- The lighting designer worked with a custom fixture manufacturer on the final shapes, layout, LED modules, and control methods to ensure that the final lighting system would work as envisioned. See Figure 1.9.

Figure 1.9 Final installation of meandering river inspired conceptual design. Lighting Design by Studio T+L, Photo courtesy of Paul Bartlett

References

Ellis, George (2007). *Let the Sun Shine In.* Supermarket News. Available at: https://www.supermarketnews.com/retail-amp-financial/let-sun-shine [accessed January 22, 2024].

Figueiro, M., Pedler, D., Plitnick, B., Zecena, E., and Leahy, S. (2023), Tailored lighting intervention (TLI) for improving sleep-wake cycles in older adults living with dementia. *Frontiers in Physiology.* 14:1290678. doi:10.3389/fphys.2023.1290678

Figueiro M., Rea M. (2016). Office lighting and personal light exposures in two seasons: Impact on sleep and mood. *Lighting Research & Technology.* 2016;48(3):352-364. doi:10.1177/1477153514564098

IES Light and Design Committee (2020). *ANSI/IES LP-1-20 Lighting Practice: Designing Quality Lighting for People and Buildings.* New York: Illuminating Engineering Society.

IES Retail Lighting Committee (2020). *ANSI/IES RP-2-20 Recommended Practice: Lighting Retail Spaces.* New York: Illuminating Engineering Society.

Heschong, Lisa (2003). *Daylight and Retail Sales*. California Energy Commission. Available at: https://newbuildings.org/wp-content/uploads/2015/11/A-5_Daylgt_Retail_2.3.71.pdf [accessed January 22, 2004].

Heschong, L., Wright, R., Okura, S. (2013). Daylighting Impacts on Human Performance in School. *Journal of the Illuminating Engineering Society*. 31. 101-114. 10.1080/00994480.2002.10748396.

Kelly, Richard (1952). Lighting as an Integral Part of Architecture. *College Art Journal*: 12:1, 24-30.

Kelly, Richard (1955). Lighting's Role in Architecture. *Architectural Forum* (February): 152, 169.

Livingston, Jason (2014). *Designing with Light: The Art, Science, and Practice of Architectural Lighting*. Hoboken: John Wiley & Sons, Inc.

Livingston, Jason (2022). *Designing with Light: The Art, Science, and Practice of Architectural Lighting, 2nd Ed.* Hoboken: John Wiley & Sons, Inc.

Sleegers PJC, Moolenaar NM, Galetzka M, Pruyn A, Sarroukh BE, van der Zande B. Lighting affects students' concentration positively: Findings from three Dutch studies. *Lighting Research and Technology* 2013;45:159-175.

Veitch, Jennifer (2018). Lighting for Productive Workplaces. In D. J Clements-Croome (Ed.), *Creating the Productive Workplace* (3rd Ed.). London: Taylor & Francis.

Veitch, J.A., Newsham, G.R., Jones, C.C., Arsenault, C.D., Mancini, S. (2010). High-Quality Lighting: Energy-Efficiency That Enhances Employee Well-Being. CIE 2010 Lighting Quality and Energy Efficiency. Vienna Austria, March 14-107, 2010, pp. 197-204.

2

Light and Vision

Abstract

To understand lighting design's influence on the perception of spaces, practitioners need an understanding of light and human vision. This chapter begins by introducing the electromagnetic spectrum in general and, more specifically the region of visible light. It reviews the three types of reflection (specular, semi-specular, and matte), as well as refraction and diffusion of light. The path of light through the human eye is discussed, as are factors affecting visual performance. The chapter concludes with a description of the three types of vision based on brightness: scotopic, mesopic, and photopic vision.

Keywords: Electromagnetic spectrum, wavelength, visible spectrum, reflection, refraction, diffusion, specular reflection, semi-specular reflection, matte reflection, angle of incidence, angle of reflection, Hunt effect, visual performance, illuminance, adaptation, contrast, contrast ratio, scotopic vision, mesopic vision, photopic vision, rods, cones, nanometer.

2.1 Introduction

Understanding light and vision is the foundation of lighting design. It is important for lighting practitioners to understand what light is, how it behaves, and how to manipulate it to achieve a desired outcome. It is also important to understand how light is perceived and the factors that affect human vision because they in turn affect, among other things, perception of contrast and color, as well as human health. This chapter provides an overview of light and vision. Light and human health are discussed in Chapter 10. Readers interested in a deeper exploration of these topics should begin by consulting the references at the end of this chapter.

2.2 The Electromagnetic Spectrum

Light can be described as that portion of the electromagnetic spectrum that produces a sensation in the human visual system. The electromagnetic spectrum is a continuous range of energies, from gamma rays at one end to radio waves at the other, that can travel through space without the need for a medium (unlike sound, for example, which needs air, water, or a solid to conduct it). If one envisions the electromagnetic spectrum as simple two-dimensional waves (see Figure 2.1), the key measurement is the distance from the crest of one wave to the crest of the next, which is called the wavelength. The wavelength range is vast, with radio waves being miles (kilometers) long, while gamma rays are 10^{-11} m or shorter. Shorter wavelengths have more energy than long wavelengths.

The visible spectrum is defined as electromagnetic energy having wavelengths of between 380 and 780 nm (nanometer or 0.000000001 meter) and makes up less than 1% of the electromagnetic spectrum.

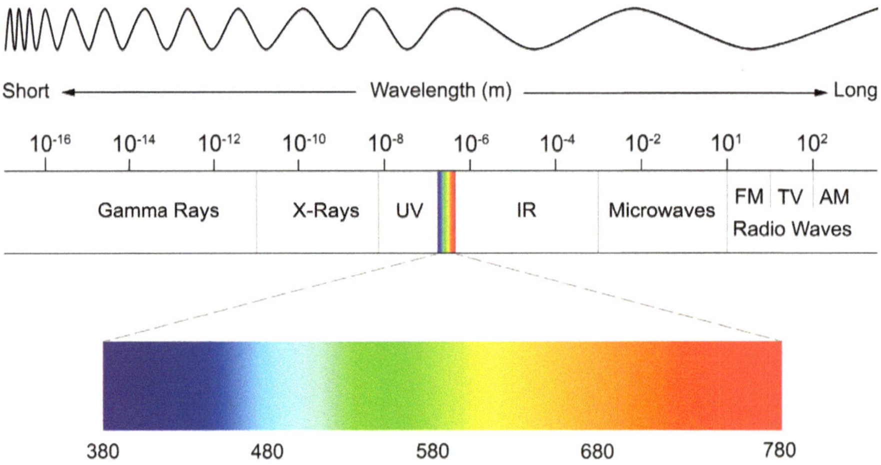

Figure 2.1 Electromagnetic spectrum.

2.3 Reflection, Refraction, and Diffusion

When light interacts with a material one of three results are possible: reflection, refraction, or diffusion.

2.3.1 Reflection

Light travels in a straight line until it interacts with a surface. Reflection describes light bouncing off a surface, as opposed to being absorbed by that surface or passing through it. There are three general types of reflection, and all three may be options for the reflector and/or trim on a luminaire.

2.3.1.1 Specular

Specular reflection is the reflection we see in a mirror and specular surfaces are smooth and mirror-like. With specular reflection, the light's angle of incidence (the angle at which light strikes a surface) is equal to the angle of exitance (the angle at which light leaves a surface) with the angle of exitance directly opposite to the angle of incidence (see Figure 2.2). Specular materials include mirrors, polished stone, and the reflectors of some luminaires. A specular reflector is the most efficient, but well-shielded recessed luminaires with specular reflectors may have visibly sharp-edged beams where the beam strikes a surface, and the reflectors may appear as dark holes in a light-colored ceiling.

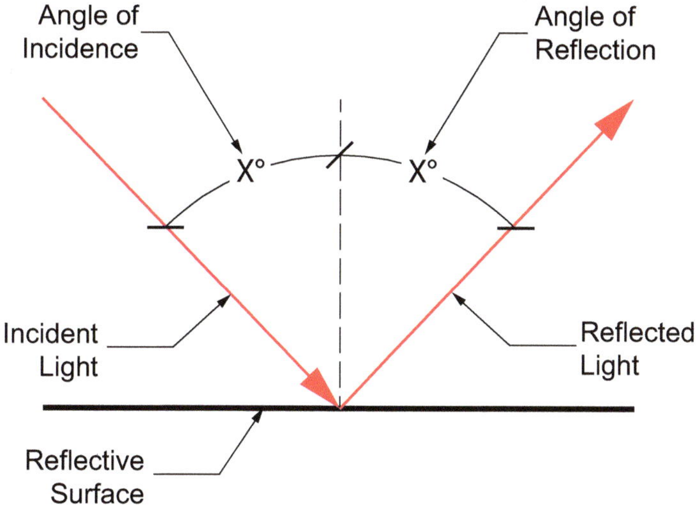

Figure 2.2 Specular reflection.

2.3.1.2 Semi-specular

Semi-specular reflection is one in which the reflected light is somewhat scattered, resulting in a cone of reflected light, as shown in Figure 2.3. Images

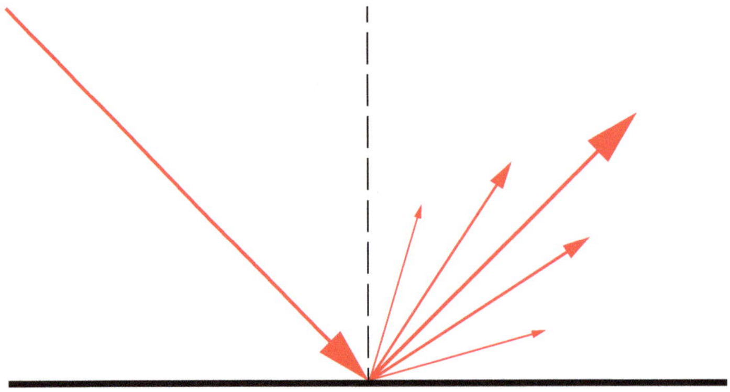

Figure 2.3 Semi-specular reflection.

seen in a semi-specular surface are softened or hazy, as is the edge of the reflected beam. Semi-specular materials include polished wood with a satin finish, high-sheen fabrics, and the reflectors of some luminaires. A semi-specular reflector is less efficient than a specular one, absorbing more light and scattering some light at higher angles rather than directing it toward the target. This may be offset by the beam having a softer edge and the reflector having a soft glow when seen in the ceiling.

2.3.1.3 Matte

Matte reflections scatter light equally in all directions, as shown in Figure 2.4. Most surfaces in our environment, from painted walls to tree leaves to the faces of other people, are matte. Matte reflectors are the least

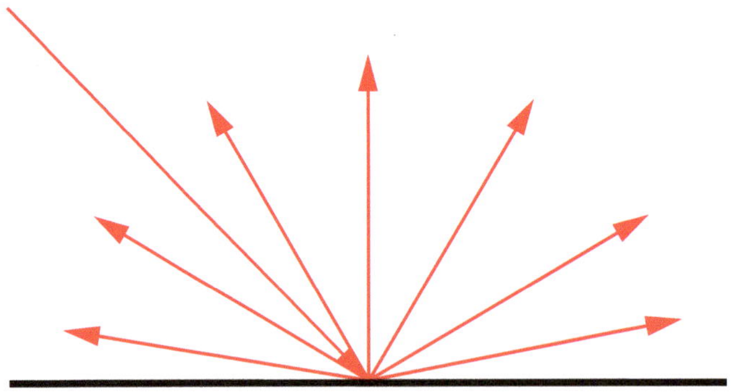

Figure 2.4 Matte reflection.

efficient, both absorbing more light and scattering more light at high angles, than semi-specular reflectors. Luminaires with matte reflectors are also the most likely to be sources of glare, so they should be selected and used with care.

2.3.2 Refraction

Refraction is the deflecting of light as it passes from a material of one density to that of another density. The design of lenses, for instance, considers the angle of incidence and the index of refraction (the amount that a given material will deflect the light as it passes from air to the material and back to air again) (see Figure 2.5).

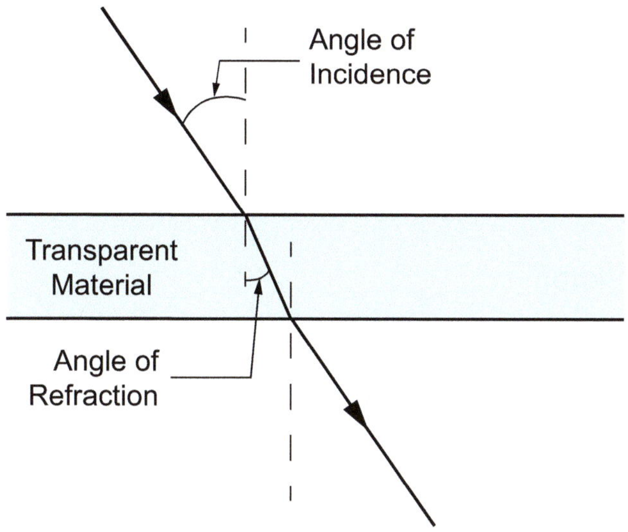

Figure 2.5 Refraction.

2.3.3 Diffusion

Diffusion is similar to refraction because diffusing materials redirect the path of the light. Unlike refractors, which generally are used to precisely redirect light, diffusers are used to scatter light, as shown in Figure 2.6. Diffusers are often used to soften the light from small sources, as with the shades on table and floor lamps. When used in this way they also behave as a larger light source and soften any resulting shadows.

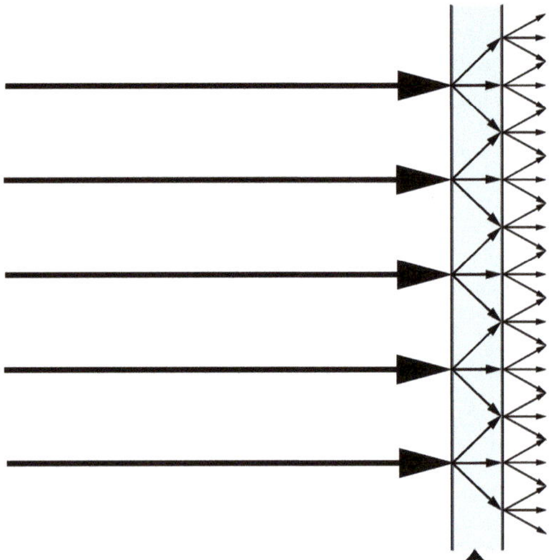

Figure 2.6 Diffusion.

2.4 Human Vision

Human vision is the result of a combination of the eye and the brain. The eye (Figure 2.7) receives light (which is moderated by the iris to adjust for

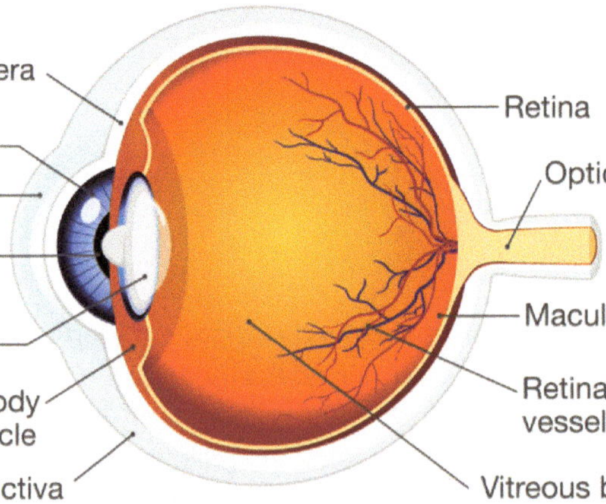

Figure 2.7 Parts of the human eye.
iStock.com/solar22

brightness) and focuses it (first by the cornea and then by the lens) on light-sensitive cells in the retina. Those cells (rods which provide black and white vision at low light levels, and three types of cones which provide color vision at typical interior and exterior daytime light levels) generate electrical signals which are then processed by other cells in the eye. The resulting signal is sent, via the optic nerve, to the visual cortex in the brain where it is interpreted, and a human observer "sees."

There are several interesting aspects of vision that are relevant to lighting designers. One of these is known as the Hunt effect, where as illumination levels decrease so does our perception of colorfulness. In other words, at low light levels colors appear more muted than they do under high light levels. The Hunt effect can have a negative impact on the appearance of objects and materials in dimly lit spaces such as spas, restaurants, and residences. Lighting designers can work to counter the Hunt effect by selecting light sources that increase vividness as measured by ANSI/IES TM-30 (see Chapter 3) (IES Color Committee, 2020).

2.5 Factors Affecting Visual Performance

There are several factors that affect visual performance, or our ability to perceive and identify objects and perform visual tasks quickly and accurately. Understanding these factors helps lighting practitioners correctly address complaints from users in existing spaces and prevent complaints in new spaces.

2.5.1 Size

Most people intuitively understand that the size of objects, or more accurately their visual angle, has a direct impact on visual performance. At the same distance, large objects are easier to see and understand than small objects. Seeing small objects or tasks, such as threading a needle, can be aided by increasing illuminance (the amount of light falling on the object) to some extent, but increasing object size or providing magnification is often a better solution.

2.5.2 Duration

Duration is another obvious factor affecting vision. It is easier to observe details when an object seen for one minute than when seen for one second. As with size, increasing illuminance can improve visual performance to a point, but increasing viewing time is often a better solution.

2.5.3 Intensity

Increasing the intensity or brightness of the light, or more accurately increasing the illuminance on the visual target, improves visual performance so long as the illuminance remains within the range of adaptation. Here's what that means:

The human eye is capable of vision over a brightness range of a trillion to 1 (1,000,000,000,000:1). However, it can't make use of that entire range at once. The average brightness sensitivity is 1,000:1, although in most applications the desired contrast range is limited to less than 4:1. The visual system's ability to adjust to different levels of brightness is called adaptation. Illumination brighter than the level one is adapted to may be experienced as glare. Illumination below the level one is adapted to may be experienced as darkness. While the solution is somewhat application dependant, if it seems that low illumination is a problem, a good approach is to combine increasing intensity with increasing contrast.

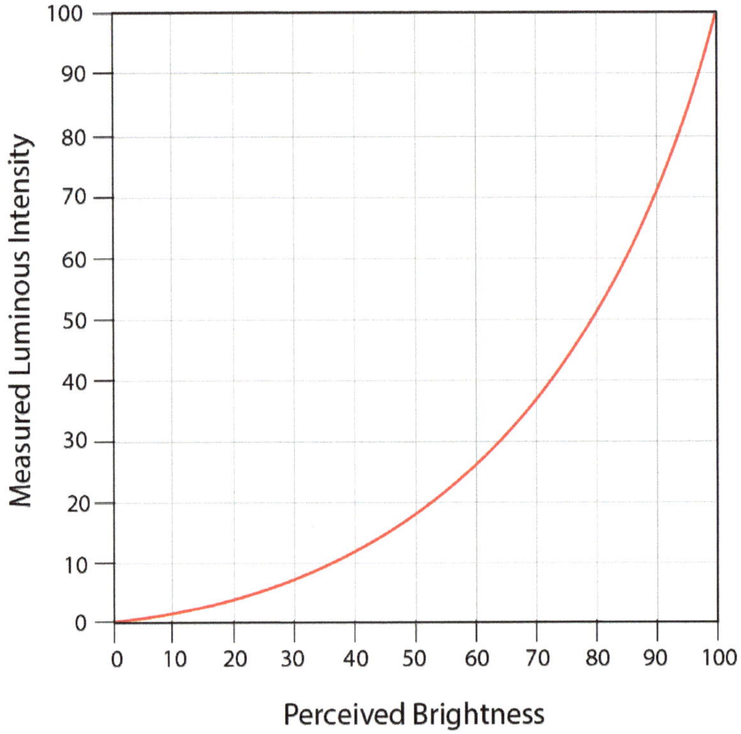

Figure 2.8 Measured intensity vs. perceived brightness.

A related issue is the perception of changing intensity. One would assume that a luminaire dimmed to 10% of full output would be perceived as 10% as bright. However, measured changes in brightness do not match perceived changes in brightness. As light intensity changes the human visual system automatically adjusts - the iris opens and closes, and the retina becomes more sensitive as light levels decline - to maintain comfortable vision over the widest range of brightness. Figure 2.8 shows the relationship between measured intensity and perceived brightness. For example, if a room is dimmed to 10% of full brightness, as measured by a light meter, it is perceived as being dimmed to 35% of full brightness. This impacts the practitioner's selection of dimming drivers for LED luminaires, which is discussed further in Chapter 5.

2.5.4 Contrast

Contrast is the relationship between the object being viewed and its immediately visible surroundings. For three-dimensional objects it may also be the relationship between the surfaces that are in higher illumination and the surfaces that are in lower illumination. A higher difference in illumination of surfaces and/or the reflectance of surfaces results in higher contrast and may improve visual performance. Likewise, color difference between an object and its background can increase contrast and improve visual performance.

While absolute contrast can be measured, those numbers need to be understood within the context of the viewer's brightness adaptation. For example, a visual target illuminated to 20 fc (200 lx) with a background of 10 fc (100 lx) is simultaneously 10 fc (100 lx) brighter than the background and illuminated at a 2:1 ratio, which is noticeable. A target illuminated to 100 fc (1000 lx) with a background at 90 fc (900 lx) is still 10 fc (100 lx) brighter than the background but is no longer twice as bright. The illuminance ratio of 1:0.9 is not discernible if one isn't looking for it.

This explains why lighting practitioners typically speak of ratios compared to the space's average illuminance. For example, The Society of Light and Lighting in the UK recommends that the work plane (the plane at desktop height) in open office plans be illuminated to 300 lx (30 fc) and that illuminance uniformity (minimum illuminance/average illuminance) across any individual work surface, such as a desk, be no more than 0.6 (Ruffles, 2018). This means that if the work plane is illuminated to 300 lx (30 fc) the darkest area on any desk should be 180 lx (18 fc). If the work plane is illuminated to 500 lx (50 fc) the darkest area on a desk should be 300 lx (30 fc).

2.5.5 Age

Age affects observers, beginning around the age of 40, in several ways. The first is that the lens yellows as people age. The yellowing of the lens blocks blue light, which is a factor in maintaining circadian entrainment (keeping the body clock synchronized to patterns of day and night, see Chapter 10) and sleep quality (IES Lighting for the Aged and Partially Sighted Committee, 2020). The lens also clouds, which scatters light within the eye and results in an increased sensitivity to glare, particularly disability glare (IES Vision Committee, 2020). Clouding also reduces the amount of light striking the retina. As a result, lighting designs for facilities for older adults should include several adjustments from typical practice to accommodate these, and other, vision changes. Adjustments should include increasing target illumination levels, reducing contrast, and selecting luminaires that provide a high level of glare control.

2.6 Scotopic, Mesopic, and Photopic Vision

The visual system performs differently depending, partially, on the amount of light the eye receives. There are three categories of light intensity: scotopic, mesopic, and photopic (IES Vision Science Committee, 2020).

Scotopic vision occurs at light levels of 0.0001 footcandles (fc) or 0.001 lux (lx) or less. This roughly corresponds to starlight and natural sky glow. At these very low levels the cones are not sufficiently stimulated, resulting in black-and-white vision from rods in the retina and low-detail resolution. This is not an illumination level that concerns lighting designers, as all designs for interior and exterior illumination are at higher levels.

Mesopic vision occurs at light levels between 1 fc and 0.0003 fc (12 to 0.004 lx). At these levels both the rods and cones are active. These illuminance levels are found between typical roadway illumination and the light of a quarter moon. There are very few interior applications where a designer would select an illumination target in this range.

Photopic vision occurs at light levels over 3 fc (30 lx) and has no upper limit. At these levels vision is dominated by the cone receptors and people perceive both color and fine detail. This is the light level of nearly all interior spaces, although very few applications call for more than 100 fc (1,000 lx).

References

IES Color Committee (2020). *ANSI/IES LS-5-20 Lighting Science: Color.* New York: Illuminating Engineering Society.

IES Light and Design Committee (2020). *ANSI/IES LP-1-20 Lighting Practice: Designing Quality Lighting for People and Buildings.* New York: Illuminating Engineering Society.

IES Lighting for the Aged and Partially Sighted Committee (2020). *ANSI/IES RP-28-20 Recommended Practice: Lighting and the Visual Environment for Older Adults and the Visually Impaired.* New York: Illuminating Engineering Society.

IES Vision Science Committee (2020). *ANSI/IES LS-7-20 Lighting Science: Vision – Eye and Brain.* New York: Illuminating Engineering Society.

Livingston, Jason (2022). *Designing with Light: The Art, Science, and Practice of Architectural Lighting, 2nd Ed.* Hoboken: John Wiley & Sons, Inc.

Ruffles, Paul, Ed. (2018). *The SSL Lighting Handbook.* London: The Society of Light and Lighting.

3

Light and Color

Abstract

The spectrum of a light affects how the colors of objects and materials are perceived, which is called color rendering. This chapter introduces and illustrates the science behind color rendering and its application to lighting design. It begins with an introduction to spectral power distribution (SPD) and chromaticity diagrams, then goes on to explain measurements of color temperature, correlated color temperature, and Duv. MacAdam ellipses and manufacturer's binning of LEDs is presented. Finally, color rendering is explained, along with the two systems for measuring color rendering, CRI and TM-30.

Keywords: Color rendering, CRI, CCT, CT, color temperature, correlated color temperature, Duv, spectral power distribution, SPD, chromaticity diagram, TM-30, R_a, R_f, R_g, $R_{cs,h1}$, $R_{f,h1}$, design intent, fidelity, vividness, preference, priority level, hue, chroma, blackbody, blackbody locus, Planckian locus, MacAdam ellipse, binning, hue shift, chroma shift, color vector graphic.

3.1 Introduction

One of the most important properties of light is that it gives the world color. Unfortunately, the light produced by the various lighting technologies tend to reveal color differently, and even two lights using the same technology can produce significantly different color appearance results. This chapter reviews some of the fundamental science of color and color rendering, followed by explanations of the two broadly accepted color rendering measurement systems: CRI and TM-30.

3.2 Spectral Power Distribution

All calculations related to light and color begin with the spectral power distribution (SPD) of the light source. The SPD is the power at each wavelength across the visible spectrum, from 380 nm to 780 nm, and is typically

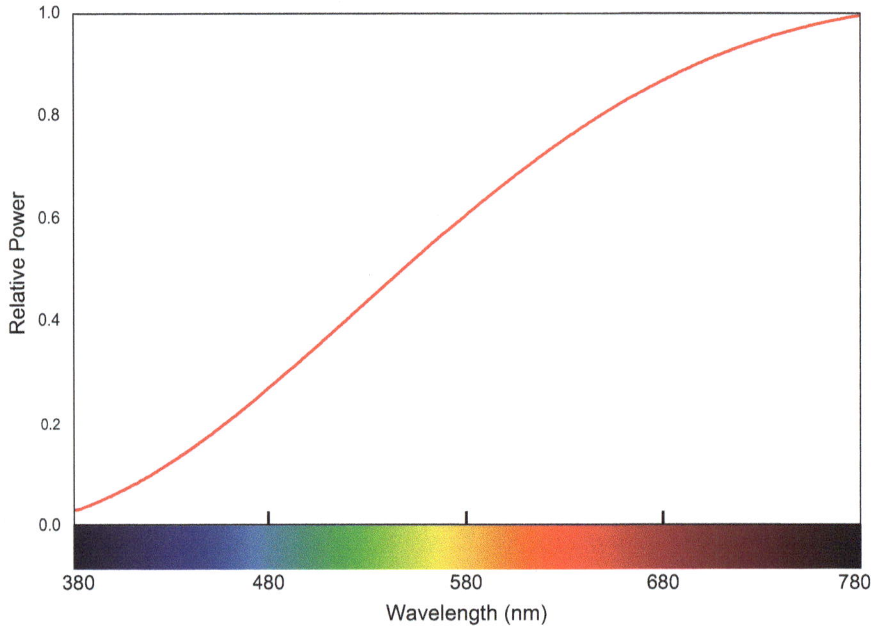

Figure 3.1 Spectral power distribution for an incandescent light source.

illustrated in a graph where the *x*-axis is the wavelength and the *y*-axis is the power (Figure 3.1), either in absolute units such as radiant watts, or in proportional units. Calculations for the measurements described in this chapter are based on a light source's SPD.

3.3 Chromaticity Diagrams

Color is a multi-dimensional phenomenon. The color of objects can be illustrated as a three-dimensional globe with axes of hue (color name), saturation or chroma (vividness or purity of the color), and lightness (a comparison of the color to the black/gray/white gradient). Color systems like Munsell use hue, chroma, and lightness to identify individual colors. It is important to note that color spaces for objects and light are not the same and cannot be used interchangeably (IES Color Committee, 2021).

In light, colors are specified by their chromaticity. The human visual system's perception of color changes with the brightness of light. To simplify the process of identifying specific colors of light the brightness is disregarded, resulting in a two-dimensional illustration called a chromaticity diagram. Chromaticity is the color of a light source independent of its intensity and is provided as two coordinates on a chromaticity diagram. Three such diagrams are used. Understanding which chromaticity diagram is used, and for what purpose, helps lighting practitioners understand the color information that is presented by manufacturers.

The first chromaticity diagram was developed by the CIE (Commission Internationale de l'Eclairage) in 1931 and has axes labelled x and y, so it is the CIE 1931 (x, y) chromaticity diagram, as shown in Figure 3.2.

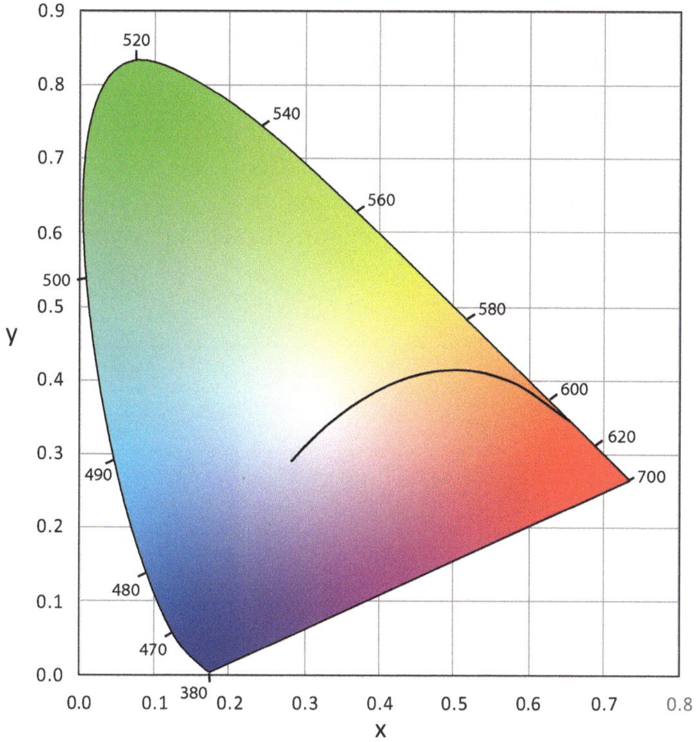

Figure 3.2 CIE 1931 (x, y) chromaticity diagram with blackbody locus.

The perimeter of the horseshoe shape is monochromatic light of a single wavelength. The line that joins the two ends of the horseshoe is called the purple line or purple boundary. Purple or magenta light doesn't exist in the linear spectrum (shown in Figure 2.1) but is perceived when red and blue light are combined. The inside of this shape is meant to contain all colors that are visible to humans, including the range of white light. Any two light sources with the same chromaticity coordinates will appear the same at equal brightness, even if there are differences in their spectra. The arc that extends into the middle of the diagram is the blackbody or Planckian locus and is the color path a blackbody radiator creates as it is heated. Blackbody radiators are explained in the next section.

In the 1940s, it was discovered that the CIE 1931 chromaticity diagram was not perceptually uniform (MacAdam, 1942), meaning that the measured distance between two points on the diagram does not correspond to the perceived difference between those two chromaticities. The 1931 diagram was replaced with a more accurate diagram in 1960, although the 1931 diagram was never withdrawn and remains the most commonly used diagram in the lighting industry. However, because it is not perceptually uniform it should be viewed with caution, especially when it is used to illustrate the color range capabilities of color-changing luminaires.

The new diagram, with axes *u* and *v* is the CIE 1960 (*u, v*) chromaticity diagram, shown in Figure 3.3. This is the chromaticity diagram where correlated color temperature (discussed later in this chapter) and some other color calculations are made.

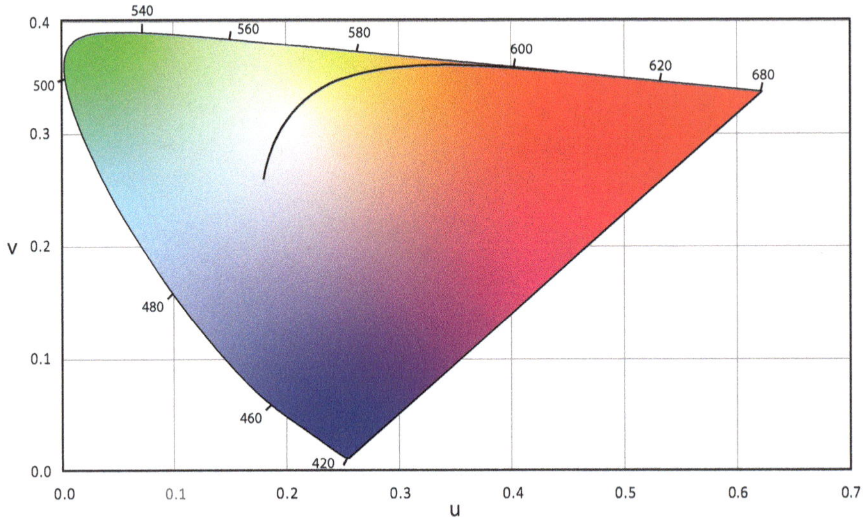

Figure 3.3 CIE 1960 (*u, v*) chromaticity diagram with blackbody locus.

Later, it was determined that the CIE 1960 chromaticity diagram was also not perceptually uniform. To correct this the *v*-axis was scaled by 1.5. The resulting diagram, shown in Figure 3.4, is the CIE 1976 (*u′, v′*) chromaticity diagram.

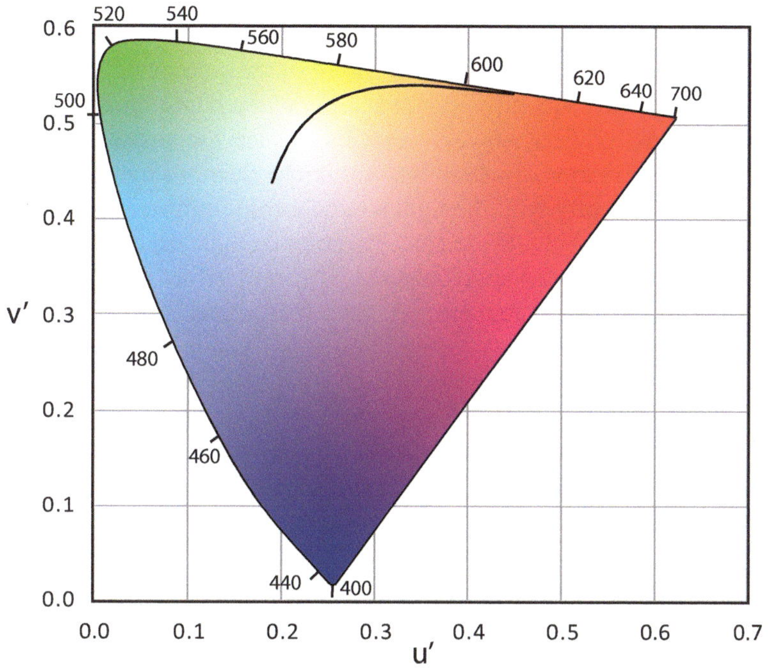

Figure 3.4 CIE 1976 (u', v') chromaticity diagram with blackbody locus.

When the CIE 1976 diagram was published the CIE simultaneously withdrew the CIE 1960 diagram, even though it is where correlated color temperature and some other color metrics are calculated. To get around this problem, and because the only change between CIE 1960 and CIE 1976 was to scale the v -axis by 1.5 to arrive at v', correlated color temperature calculations are technically made in CIE 1976 (u', $2/3\ v'$) which reverses the scaling to return to CIE 1960. For clarity, the rest of this chapter will continue to refer to CIE 1960 and CIE 1976.

3.4 Color Metrics

Lighting practitioners primarily work with white light. However, there are many versions of white light, so there are several metrics related to white light that are meaningful. They include a measurement of the warmth or coolness of white light (color temperature and correlated color temperature), the pinkness or greenness of the light (Duv), and how well a given light source enables us to see the full range of object and material colors without distortion (color rendering).

3.4.1 Color temperature

In discussions with clients, lighting practitioners may describe light with widely understood terms such as warm and cool. When communicating with others in the lighting industry, however, those terms are too vague. There are far more than two variations of white light to choose from, so the options are described more specifically with the light source's color temperature (CT).

Everyone is familiar with the idea that the color of light emitted by heated objects changes as the temperature changes. Light emitted by a piece of metal, for instance, begins as red at low temperatures then passes through orange, yellow and becomes white as its temperature rises. Color temperature is a way to specifically identify the color of light along this warm/cool axis. It does so by comparing the color of light emitted by a light source to the light emitted by a blackbody radiator. A blackbody radiator is a theoretical object that is a perfect absorber and emitter of all electromagnetic energy. The spectrum of light emitted by the blackbody radiator is solely determined by its temperature, measured in kelvins (K) (IES Color Committee, 2021). Note that the blackbody radiator is also called the Planckian radiator after Max Planck who, in 1900, devised the formula to describe it. SPDs of the blackbody radiator at several color temperatures are shown in Figure 3.5.

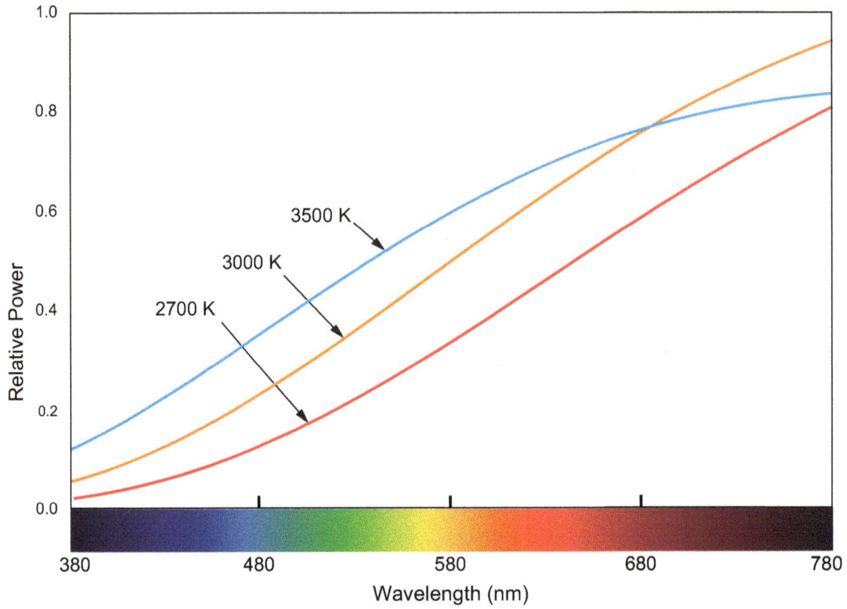

Figure 3.5 SPDs of a blackbody radiator at three color temperatures.

3.4.1.1 Correlated color temperature

Color temperature only applies to SPDs exactly matching that of a blackbody radiator which, for all practical purposes, means incandescent lamps. The SPD of other light sources can vary dramatically from the blackbody radiator, as shown in Figure 3.6. Correlated color temperature (CCT) is used to describe the warmth/coolness of non-incandescent light sources. CCT is defined as "the absolute temperature of a blackbody whose chromaticity most nearly resembles that of the light source" (ANSI/IES LS-1-22, 2022). All non-incandescent sources have a CCT, not a CT, although generally in professional practice the two terms are interchangeable.

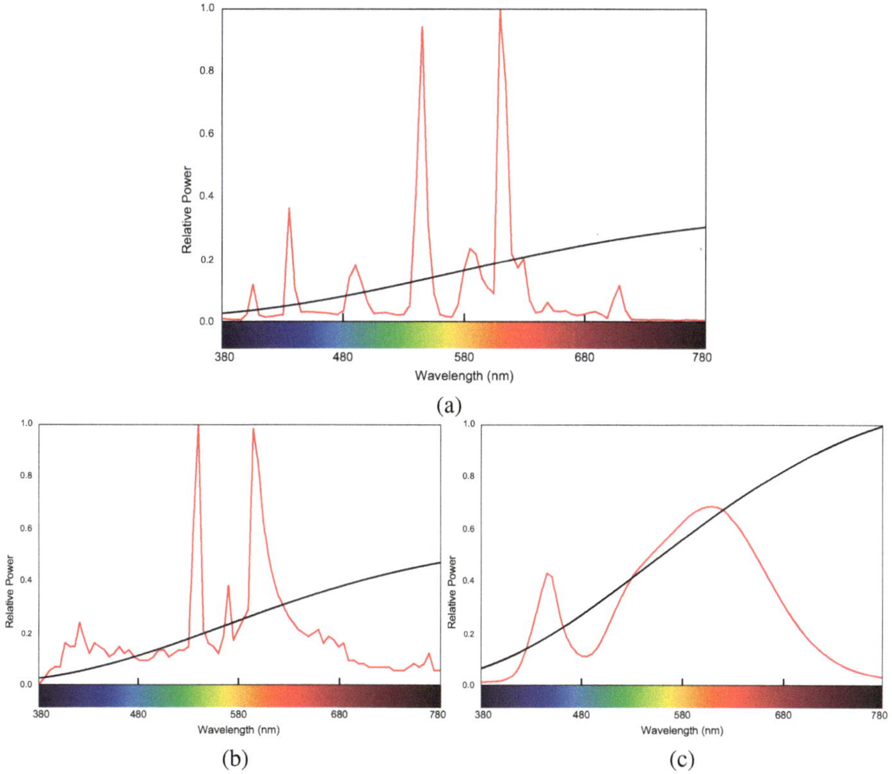

Figure 3.6 SPD of the blackbody radiator and other light sources. (a) Fluorescent lamp at 3000 K (b) metal halide lamp at 3000 K (c) LED lamp at 3000 K.

The color temperatures available for LEDs are the same as those originally established in 1956 for fluorescent lamps (IEC 60081:1956, now withdrawn and replaced by IEC 60081:1997 and ANSI/NEMA C78.376) but

Table 3.1 Nominal LED correlated color temperatures used in general lighting.

Nominal LED correlated color temperature (K)	**Description**
2700	Warm white light that is warmer than standard incandescent (the CT of a standard 100 W incandescent lamp is 2856 K)
3000	Warm white that is slightly whiter or cooler than a standard incandescent
3500	Neutral white light
4000	Moderately cool white light
5000	Cool white light that may be described by some as daylight (daylight color temperature varies but is usually described as 5000 and above)

are defined by a different standard, called ANSI/NEMA C78.377. They are described in Table 3.1.

3.4.2 Duv

Color temperature can be thought of as an axis from amber (low CCT) to blue (high CCT) but color is multi-dimensional. CCT does not identify a single chromaticity, and some white light sources can appear warmer or cooler and/or greener or pinker than others, even if they have the same CCT. The green/pink aspect of color appearance can be thought of as an axis perpendicular to the amber/blue axis and is called Duv.

Duv stands for the distance from the blackbody locus measured in the CIE 1960 (u, v) chromaticity diagram. Duv provides information about the direction and distance from the blackbody locus, and the resulting color appearance of light. A light source with a positive Duv plots above the blackbody locus on a chromaticity diagram and has a yellow or green tint, depending on CCT. A light source with a negative Duv plots below the blackbody locus on a chromaticity diagram and has a pink or lavender tint, depending on CCT. To understand the magnitude of the chromaticity shifts, a Duv of 0.001 is approximately equal to a CCT shift of 50 K and most commercially available white light LEDs have a Duv of between -0.007 and 0.005 (Livingston *et al.*, 2021).

Combining CCT and Duv gives a lighting practitioner a good idea of the color appearance of the white light in question. Very few luminaire manufacturers provide Duv in their documentation so practitioners who are interested must request it. Fortunately, Duv, like CCT, is calculated from the luminaire's SPD. Therefore, if the manufacturer doesn't have a record of the Duv, it should be easy for them to calculate.

3.4.2.1 Duv and efficacy

The human visual system is not equally sensitive to all wavelengths of light. Human vision is most sensitive to light with a wavelength of 555 nm (greenish-yellow) and sensitivity decreases toward either end of the visible spectrum. This sensitivity function is called the photopic luminous efficiency function and is illustrated in Figure 3.7. The symbol is $V(\lambda)$ (vee lambda), which also how it is usually referred to.

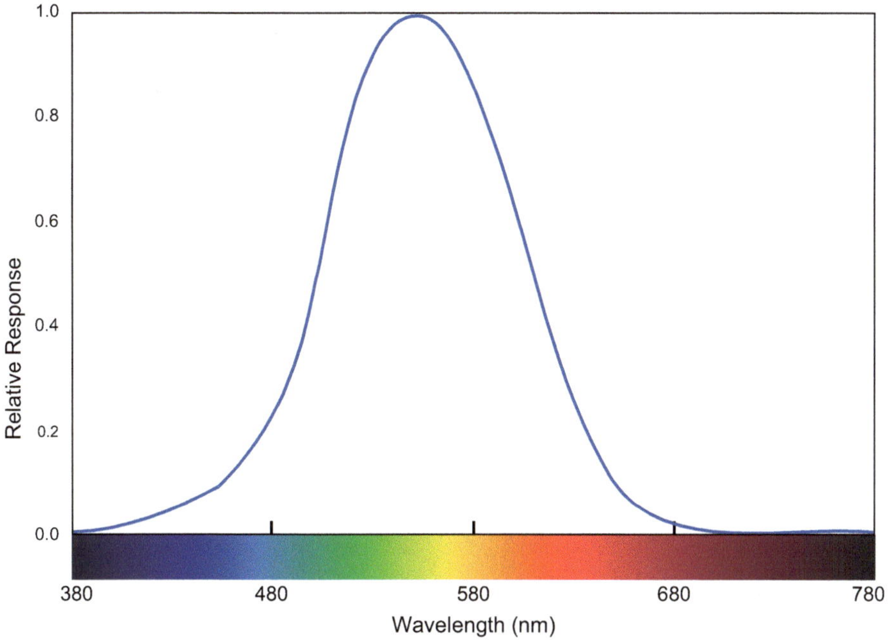

Figure 3.7 $V(\lambda)$.

One part of the definition of the lumen includes weighting the contribution of each wavelength according to $V(\lambda)$, which means that deep red and blue light contribute relatively little to brightness, while yellow-green makes a large contribution. Therefore, the most efficacious light source would be one that produces only light with a wavelength of 555 nm. Such a light source would have an efficacy of 683 lm/W (IES Color Committee, 2021) but would be unusable because no one would accept an environment illuminated with yellow-green light for any application.

One result of the use of $V(\lambda)$ is that light source manufacturers must balance efficacy with other aspects of light such as color quality – that is maximizing the amount of light near 555 nm to increase efficacy while still providing enough light in the rest of the spectrum to produce acceptable, or even pleasing, white light. Light sources that appear to be too green, and therefore have a large positive Duv (above 0.003, for example), are likely to be the result of a manufacturer's over emphasis on efficacy to the detriment of color quality. When in doubt, practitioners can obtain sample lamps or luminaires to make an in-person evaluation of the light, and compare their observations to the manufacturer's efficacy, color rendering, and other metrics.

3.4.3 LED correlated color temperature

As previously explained, LEDs of the same CCT do not necessarily produce light of the same color. CCT and Duv are continuous ranges, not single points on a chromaticity diagram. This leads to the question, "Who defines LED white light CCT and how?" The answer is the National Electrical Manufacturers Association (NEMA) in a document titled *ANSI/NEMA C78.377 American National Standard for Electric Lamps – Specifications for the Chromaticity of Solid-State Lighting Products.*

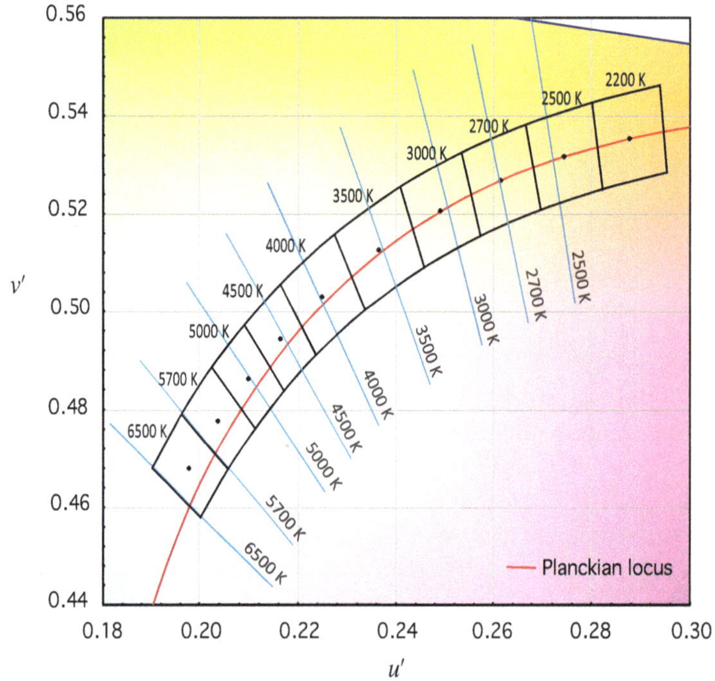

Figure 3.8 ANSI/NEMA C78.377 CCT quadrangles and center points in the CIE 1976 (u', v') chromaticity diagram.

C78.377 defines the four corner points and center points of "chromaticity quadrangles" for 10 CCTs and a tuneable white range of 2300–6400 K, as shown in Figure 3.8. The quadrangles are defined in both CIE 1931 and CIE 1976 chromaticity diagrams. Any LED producing light that falls within a quadrangle is said to produce light of that nominal CCT. This helps manufacturers minimize waste that might otherwise be created by the variations that occur in LED production. The quadrangles are quite large, though, (seven-step ellipses, which are discussed later in this chapter) and LEDs from different regions of a single quadrangle will produce noticeably different colors of white light even though they have the same nominal CCT. To control the variation in what LEDs are sold to customers, manufacturers sort their LEDs in a process called *binning*, which is described next.

3.4.4 LED binning

As mentioned in Section 3.4.3, the chromaticity quadrangles of C76.377 are quite large, and an LED that falls into one corner will be noticeably different from one in another corner even though they have the same nominal CCT. To manage the variability inherent in LED manufacturing, each LED is tested for several characteristics, including its chromaticity, and LEDs with similar characteristics are grouped into "bins." Binning by LED manufacturers lets luminaire manufacturers be more selective in the LEDs they purchase by choosing a set of characteristics that are more limited and specific than just a given CCT. Unfortunately, bins are not standardized across the industry. Each LED manufacturer has developed their own set of characteristics and measurement ranges to sort their products.

3.4.5 MacAdam ellipses

As mentioned in Section 3.3, the CIE 1931 (x, y) chromaticity diagram is not perceptually uniform. This was discovered and explained in a paper by David MacAdam, a researcher at Eastman Kodak who was looking into visual sensitivity and color differences (MacAdam, 1942). In his experiment, a trained observer made multiple attempts to match one variable light to one of a fixed chromaticity across 25 different chromaticities. The assumption was that the point at which a difference between the two became undetectable would form a circle around the fixed point. However, as Figure 3.9 illustrates, they actually formed ovals, or ellipses, of varying size and orientation depending on the chromaticity. These ellipses are now known as MacAdam ellipses and are used as a measure of the variation that luminaire manufacturers accept in the LEDs they use.

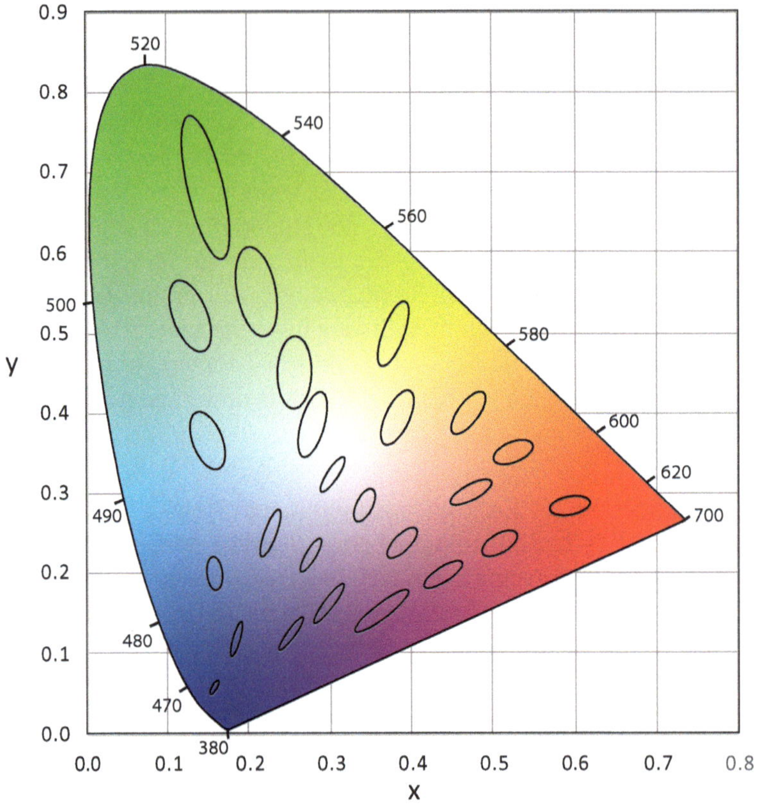

Figure 3.9 MacAdam ellipses (shown at 10× size).

Today, the distance from the center point to the edge of one of MacAdam's ellipses is a MacAdam "step," so such an ellipse would be two steps across, or a two-step MacAdam ellipse. An ellipse that is two steps from center to edge is a four-step MacAdam ellipse, etc. As Figure 3.9 shows, the size and orientation of the original 25 MacAdam ellipses is not consistent across the chromaticity diagram so ellipses for all other chromaticities must be interpolated. Unfortunately, there is not an industry-wide standardized method of interpolation – every manufacturer uses their own system.

In 2014 the CIE proposed resolving this problem by using circles in the CIE 1976 chromaticity diagram (International Commission on Illumination, 2014), noting that a circle with a radius of 0.0001 roughly corresponds to a MacAdam ellipse. This recommendation has not yet been broadly adopted by the industry and MacAdam ellipses remain the default standard.

Another term manufacturers use is standard deviation color matching (SDCM). The size if the original MacAdam ellipses is defined as one SDCM. Therefore, the size of 1 SDCM varies with chromaticity, just as they do for a MacAdam ellipse (International Commission on Illumination, 2014).

Given the above, when reviewing manufacturers specification sheets lighting specifiers can use the following rules-of-thumb.

- A set of luminaires manufactured with LEDs from within a three-step ellipse or smaller will not have problems with luminaire mismatch. There may be slight differences between luminaires, but most people in most applications will not notice them.
- Luminaires using a four-step ellipse may have color differences that are noticeable but will likely acceptable for most people in most applications.
- A five-step or more ellipse permits noticeable variation (Davidson and Freide, 1953) that will likely be unacceptable for most indoor applications, although it may be acceptable for storage or mechanical rooms, warehouses, and other spaces where luminaire to luminaire color consistency may not be required.

3.5 Color Rendering – CRI

The SPDs shown in Figures 3.5 and 3.6 help to illustrate an important point: the spectrum of light sources varies with the color temperature or CCT, Duv, and light source technology. Changing the spectrum of a light source will change the spectrum of light reflected from materials and objects in a room. The changes to the reflected spectrum may be unnoticeable, noticeable but minor, or even significant enough to completely change an object's color appearance. This gives rise to the problem of color rendering, or the color appearance of materials and objects when viewed under different light sources. The CIE and IES both define color rendering as "the effect of an illuminant on the color appearance of objects by conscious or subconscious comparison with their color appearance under a reference illuminant."

Note that this definition relies on a comparison. The color appearance of objects under a standardized light source is compared to that under the light source in question. In this approach a close match is "good" color rendering, and a mismatch is "poor."

Light sources with "good" color rendering minimize all color shifts so that objects appear as they are expected or known to appear. Any color shifts that occur are acceptable.

A light source with "poor" color rendering will cause color distortions that may increase or decrease the saturation of colors (making an apple appear be more gray or more vividly red than it is, for example), shift the hue of colors (by making a red apple appear more yellow/green or more purple than it is, for example), or an unacceptable combination of the two.

Color rendering wasn't an issue with incandescent lamps but became a problem with the broad adoption of fluorescent lights in the 1940s and 1950s and continues with today's LEDs. The need for a method to measure and report the influence of the spectrum on color rendering resulted in *CIE 13-1965 Method of Measuring and Specifying Colour Rendering Properties of Light Sources*, which was published in 1965. The current version is *CIE 13.3-1995*. The method is widely known as CRI because of the measure it produces, the general color rendering index, and until 2015 was the only industry-wide method of communicating color rendering information.

CRI (technically, the measure is the general color rendering index or R_a) is a fidelity measurement that is calculated by illuminating the first eight color samples shown in Figure 3.10 with the reference illuminant and the light source in question, and then comparing the reflected spectra. A calculation is made for the difference between the two spectra for each color sample, and those differences are averaged. If the resulting number is 100 it means that the two spectra are a perfect match. Any number below 100 indicates some difference – the lower the number the greater the difference. Note that no physical comparison is made – it is entirely mathematical using the SPD of the light sources and the spectral reflection function of the color samples. LEDs are designed and manufactured to have a CRI or R_a of at least 80,

Figure 3.10 The eight colors used to calculate CRI R_a (top) and six supplemental colors, including R_9 (bottom).

and many luminaire manufacturers offer LED options with two or more CRI values, often CRI in the 80s and 90s.

The single number makes CRI easy to use. As a measure of fidelity, a higher number means a closer match to the reference light source, so higher is better. However, a CRI of 80, for example, means that some colors are not rendered well. The single number tells us nothing about which colors are poorly rendered, or how. One piece of additional information is sometimes provided – R_9. Recall that the color samples are numbered 1 to 14. Color sample 9 is deep red. There is a great deal of research showing that the human visual system places an outsized importance on red in interpreting the colors seen. Supplementing R_a with R_9 provides a bit more insight into the color rendering of a light source. Note that R_9 is not scaled the same as R_a, so any positive value is considered good.

Over the years several misconceptions about CRI have taken root: high CRI means the light produced is "better," that it makes objects appear more "natural," that it shows the "true" color of objects, etc. None of these are accurate. CRI, as a fidelity metric, only indicates how closely the light source in question matches the reference light source. Depending on the application, a high-fidelity light source that renders colors very similarly to the reference illuminant may, or may not, be desirable.

As early as the 1970s some flaws in the CRI calculation became known. Also, although color and vision sciences have made advances since 1965, CRI has not been updated to include them. The result is that today CRI is known to be a measure with several errors in the calculation (IES Color Committee, 2021), including:

- Averaging the results of the eight color samples means that poor color rendering of one may be hidden by high color rendering of another.
- The single number of CRI provides very little information, meaning practitioners have no way of knowing if one or more colors will be rendered poorly.
- The eight color samples do not adequately cover the color space, and as a result light sources can be "optimized" to achieve a higher CRI value than a visual examination of the color rendering suggests.
- Several instances of outdated science are still in the calculation, including the use of the CIE 1964 UCS color space, which is not recommended for any use.
- Shifting from a reference light source of a blackbody radiator below 5000 K to a model of daylight above 5000 K can cause CRI number to change dramatically at that threshold for tunable white luminaires.

- The narrow band spectrum of LEDs produces lower CRI values than a visual examination of the color rendering suggests.

Although these flaws, and others, have been known for some time, the CIE committee responsible for managing CRI has been unable to agree on updating the calculations. In 2017 the CIE published *CIE 224:2017 Color Fidelity Index for Accurate Scientific Use*, which adopted the fidelity measure R_f of ANSI/IES TM-30 (described later in this chapter) but specifically recommended it only for scientific use, not use by the general lighting industry. The result is that CRI continues to be the dominant color rendering metric, but it should be used with caution. Practitioners who are concerned about accurate measures of color rendering, and measures beyond fidelity, should use ANSI/IES TM-30, which is discussed later in this chapter.

3.5.1 Beyond fidelity

Since CRI has been the only industry-wide measure of color rendering, color rendering and fidelity have come to mean the same thing to many practitioners. However, there are other aspects of color rendering that practitioners could take advantage of if a system that measured them existed (such a system does exist in ANSI/IES TM-30, which is described in the next section). Among them are:

- Naturalness – light renders objects as expected. While fidelity and naturalness are often associated with each other, they are separate design goals and fidelity is not the best measure to achieve naturalness.
- Preference – light renders object colors with some distortions, but those distortions are preferred over a high-fidelity light source.
- Vividness – light increases or decreases the chroma, saturation, or vividness of objects.
- Discrimination – light permits observers to detect even small differences in the colors of objects.

To understand and implement these color rendering goals, it is helpful to understand how a lights source's SPD affects the color shifts that are associated with them.

3.5.1.1 Color shifts

There are two types of color shifts that are components of color rendering: hue shifts and chroma shifts, as shown in Figures 3.11 and 3.12. Hue shifts occur when the SPD causes a change to the color of an object, making a ripe yellow banana seem to be an unripe green one, for example. Hue is an important

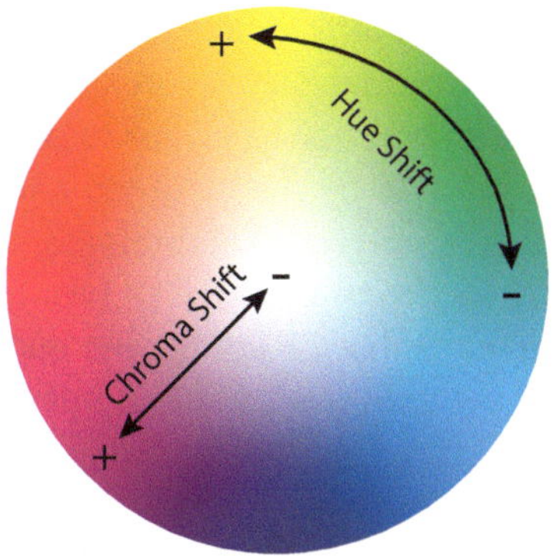

Figure 3.11 Hue and chroma shifts.

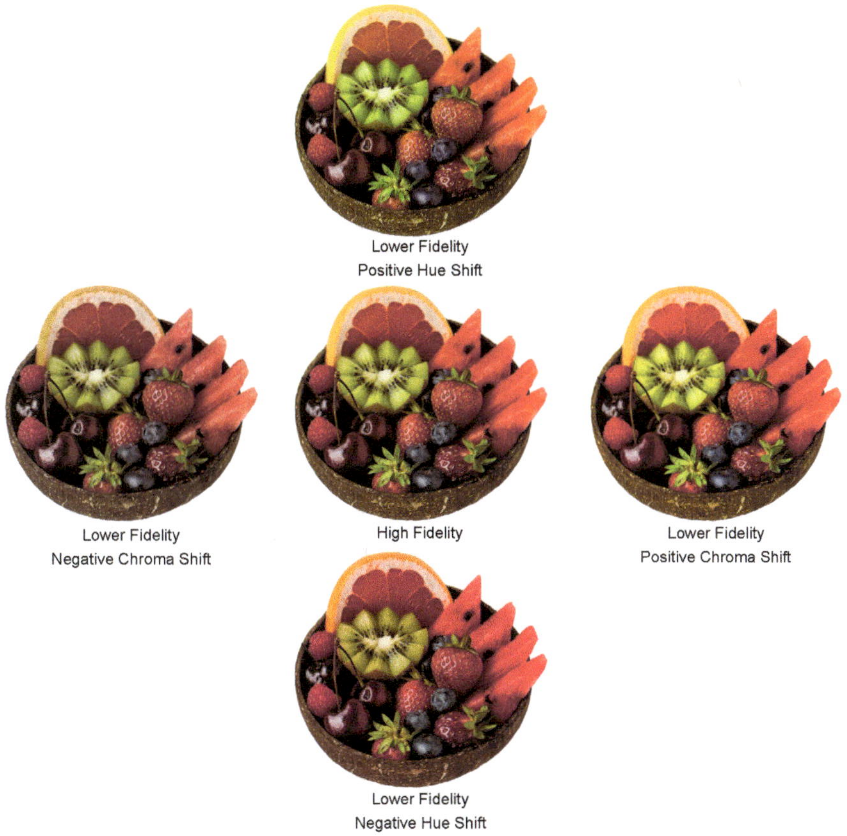

Figure 3.12 SPDs producing hue and chroma shifts.
Original photo by Jo Sonn on Unsplash

aspect of identifying and distinguishing objects. Chroma shifts occur when the SPD causes a change to the chroma, saturation, or vividness of objects. A light source that produces chroma shifts could make a fresh ripe apple look like an old rotting one, for example, by reducing the chroma. Chroma shifts play a role in perceptions of preference and naturalness, among others.

Figure 3.12 illustrates the weakness of CRI. All four of the outside photos could be caused by light sources with a CRI of 80, but that single number provides no information about how the light source in question differs from the reference source. A color rendering system that fully explored and explained the impact of the light source's spectrum would include information about:

- Color fidelity
- Chroma shifts
- Hue shifts
- The perceptual impacts of the above (preference, naturalness, vividness, etc.)

Fortunately, in 2015 a color rendering measurement system that delivers on these bullet points was introduced – ANSI/IES TM-30.

3.5.2 Color rendering – ANSI/IES TM-30

Accurately measuring the color rendition characteristics of a light source is obviously a concern of the lighting industry. More than a dozen metrics were introduced over the years to address issues of fidelity, color discrimination, preference, etc. but none of them earned widespread adoption. In 2015 the Illuminating Engineering Society combined these ideas with the most up to date color and vision science to produce *ANSI/IES TM-30-15 Technical Memorandum: IES Method for Evaluating Light Source Color Rendition*, which is usually referred to simply as TM-30 (as of 2024 the current version is TM-30-24). TM-30 overcomes CRI's weaknesses, produces 149 measures that can be used singly or in combination to evaluate a wide range of color rendering characteristics, and produces several graphics that illustrate the calculated results. It also provides research evidence for how the measures can be used, and recommended values to achieve a set of color rendering goals.

Among the many improvements over CRI, two are most immediately meaningful to lighting practitioners. The first is the use of 99 color evaluation samples (CES) that are drawn from real-world objects with reflectance of greater than 20% and are evenly distributed throughout the color space. This anchors the color rendering calculation in the objects and materials that are being illuminated and prevents lamp manufacturers from "optimizing" spectra to achieve higher scores. The second is a graduated transition from a blackbody radiator to a CIE model of daylight over the range of 4001 to 4999 K. TM-30 uses the same reference illuminants as CRI, but this transition prevents abrupt changes in calculated values that can occur with CRI, especially with tunable white luminaires.

The most important graphic TM-30 produces is the color vector graphic (CVG) which, with a few additions, is the largest component of the simple report, as shown in Figure 3.13. TM-30 also produces an intermediate report and a full report with additional detailed information.

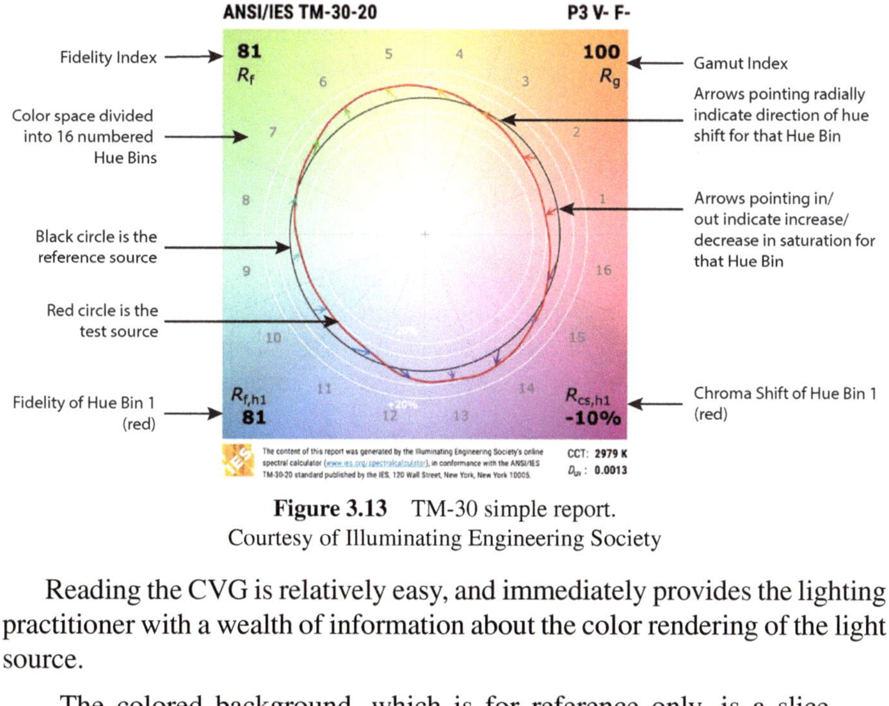

Figure 3.13 TM-30 simple report.
Courtesy of Illuminating Engineering Society

Reading the CVG is relatively easy, and immediately provides the lighting practitioner with a wealth of information about the color rendering of the light source.

> The colored background, which is for reference only, is a slice through the equator of the color space used (CIE CAM02 UCS). This is divided into 16 numbered wedges called Hue Angle Bins or Hue Bins. The center of the background is white light. As one moves toward the perimeter light takes on a color, which becomes increasingly saturated.
>
> The black circle represents the color rendering of the reference light source.

The red shape represents the color rendering of the light source as compared to the reference light source. When the red shape is inside the black circle, it indicates that the light source causes colors in that Hue Bin to decrease in saturation. When the red shape is outside of the black circle it indicates that the light source causes colors in that Hue Bin to increase in saturation. When the red shape and black circle overlap there is no change in saturation.

The arrows along the black circle start at the center of each Hue Bin. Their direction and length also indicate the direction and magnitude of chroma, or saturation, shifts when they are perpendicular to the black circle. When they point along the circle it indicates that the light source causes hue shifts, with the direction and length of the arrows indicating the direction and magnitude of the shift.

The four corners of the CVG hold the results of the four key TM-30 measures: R_f, R_g, $R_{\mathrm{cs,h1}}$, and $R_{\mathrm{f,h1}}$.

3.5.2.1 TM-30 measures

TM-30 produces 149 measures that provide a full analysis for the color rendering of a light source. While 149 measures are obviously too many to be practical, the full suite of measures has scientific usefulness. Lighting practitioners can group similar measures to simplify understanding without losing detail, as follows (Table 3.2):

Table 3.2 Summary of TM-30 measures (Royer, 2022).

Measure	Symbol	What it characterizes (relative to the reference illuminant)	Interpretation of values	Typical values
Fidelity index	R_f	Average color shift for all 99 CES	Values closer to 100 indicates closer match to the reference	70 to 100
Gamut index	R_g	Average chroma shift for all 99 CES	Values above 100 indicate average increase. Values below 100 indicate average decrease.	80 to 120

Table 3.2 *(Continued).*

Measure	Symbol	What it characterizes (relative to the reference illuminant)	Interpretation of values	Typical values
Local chroma shift	$R_{\mathrm{cs,hj}}$	Average chroma shift for CES within a hue bin	Values above 0% indicate increase. Values below 0% indicate decrease.	–20% to 20%
Local hue shift	$R_{\mathrm{hs,hj}}$	Average hue shift for CES within a hue bin	Positive values indicate counter clockwise shift. Negative values indicate clockwise shift.	–0.2 to 0.2
Local Fidelity	$R_{\mathrm{f,cesi}}$	Average fidelity for CES within a hue bin	Values closer to 100 indicates closer match to the reference	60 to 100

Notes:

1. For local chroma shift and local hue shift *j* is a placeholder for the hue bin number (1–16).
2. For local fidelity *i* is a placeholder for the CES number (1–99).

3.5.2.2 TM-30 design intents and priority levels

A 2018 analysis of several high-quality studies of TM-30 demonstrated that it can be effectively used, at illuminance levels between 200 and 700 lux (20 fc to 70 fc) (see Chapter 8), to assess aspects of color rendition beyond fidelity, and that only four measures are needed. This simultaneously expanded the lighting practitioner's color rendering design goals and simplified the use of TM-30. The color rendering goals, referred to as design intents, are Preference, Vividness, and Fidelity and are illustrated in Figure 3.14.

> The Preference (P) design intent is used when the goal is to create a pleasing, natural-looking environment. Preference permits, and even rewards, particular color shifts, especially moderate increases in red chroma. Preference may be the primary design intent in

applications such as hospitality, residential, retail, and office lighting. Note that light sources with spectra that increase the chroma of red may be less energy efficient than high-fidelity sources.

The Vividness (V) design intent is to create a vivid or vibrant environment, which at high levels of vividness may not be considered natural-looking. Light sources meeting the Vividness intent produce a large average increase in chroma with emphasis on increasing red chroma. Vividness may be the primary design intent in certain entertainment or retail applications. Note that, as with Preference, light sources with spectra that increase the chroma of red may be less energy efficient than high-fidelity sources.

The Fidelity (F) design intent is to render objects similarly to their color appearance under the reference illuminant. Fidelity may be the primary design intent in medical, manufacturing, or color reproduction applications.

(a) (b) (c)

Figure 3.14 A pizza illuminated with (a) a high-fidelity light source (b) a high-preference light source (c) a high-vividness light source.

Original image iStock.com/etorres69

Each design intent is divided into three priority levels. Priority Level 1 sets the strictest criteria and is most likely to achieve the intent for the highest percentage of viewers. The strict criteria, though, may result in fewer luminaire options, higher luminaire costs, etc. Priority Level 3 sets the loosest criteria and is least likely to achieve the design intent. This does not mean that the design intent is not achieved, only that it is successfully achieved for fewer viewers (but still a majority). The loose criteria may permit luminaires that barely achieve the design intent. Table 3.3 provides a description of each of the nine Design Intent/Priority Level combinations.

Table 3.3 TM-30 Annex E design intent and priority level descriptions.

<table>
<tr><td colspan="2" rowspan="2"></td><td colspan="3">Design intent
The desired effect of color rendition on the illuminated environment</td></tr>
<tr><td>Preference</td><td>Vividness</td><td>Fidelity</td></tr>
<tr><td rowspan="3">Priority Level
The balance between allowing for tradeoffs and increasing the likelihood of meeting the design intent</td><td>1</td><td>Highly preferred compared to their reference source. Emphasizes increased red saturation.</td><td>Creates dramatic increases in chroma for most or all hues. The increased vividness may be viewed as unnatural.</td><td>Permits only minor shifts in chroma and hue compared to the reference light source.</td></tr>
<tr><td>2</td><td>Broadly preferred compared to their reference source. Many colors are increased in saturation, with an emphasis on reds.</td><td>Enhances the average saturation of colors with a focus on enhancing reds.</td><td>Provides very good color fidelity by minimizing hue and chroma shift, in part by focusing on the fidelity of reds.</td></tr>
<tr><td>3</td><td>Generally preferred to their reference source. Some colors are increased in saturation while limiting the desaturation of reds.</td><td>Renders colors with at least the same saturation as the reference light source.</td><td>Meets the minimum requirement for fidelity by minimizing hue and chroma shifts compared to the reference light source.</td></tr>
</table>

The design intents are not mutually exclusive, and many overlaps exist. This may allow practitioners to have a primary and secondary design intent, or to use one design intent to set limits on another. The design intent and priority level are provided in the upper right corner of the TM-30 simple report, as shown in Figure 3.13. A light source failing to meet the criteria for priority level 3 is indicated with a dash, as in P1 V- F2.

The design intent and priority level shorthand makes understanding and using TM-30 much easier than using the measures and values that underly it. It is designed to be used in luminaire schedules and specifications to set color rendering requirements and assists a lighting practitioner in writing and holding a color rendering specification.

3.5.2.3 Designing with TM-30

Using CRI, regardless of the project type, a practitioner would typically say a light source with higher fidelity is better than one with a lower fidelity, which permits color shifts that can't be predicted. Using TM-30 a practitioner can have a good understanding of the color shifts caused by a light source, analyze the project's color rendering requirements and goals, and select a light source that meets those goals.

For example, during the early phase of the design for a high-end restaurant the owner may express the desire that "the diners, the food, and the interior should all be beautiful." The lighting practitioner might interpret this as meaning the color rendering should be highly preferred versus the reference, and choose P1 (Design Intent: Preference, Priority Level: 1) as the primary specification. Concerned that the maximum increased red saturation permitted in P1 might be too high for the project, the lighting practitioner may add a secondary specification of F3 (Design Intent: Fidelity, Priority Level: 3) to limit red saturation. The resulting color rendering specification would be P1 V- F3.

During subsequent phases of the design, as various luminaires are being considered, the practitioner would look for TM-30 information, often in the form of the simple report shown in Figure 3.13. The TM-30 report might be on the luminaire cut sheets or in supplemental downloads on each website. If the information is not available, the practitioner would contact the local sales representative or the manufacturer to request a TM-30 report. The reports would be part of the evaluation of each luminaire, and luminaires that didn't meet the P1 V- F3 specification, or that didn't provide a TM-30 report, would not be considered for the project.

References

ANSI/IES LS-1-22 Lighting Science: Nomenclature and Definitions for Illuminating Engineering. Available at: https://www.ies.org/standards/definitions/ [accessed December 6, 2023]

Davidson, Hugh, R., Freide, Elaine (1953). "The Size of Acceptable Color Differences," *J. Opt. Soc. Am.* 43, 581-589

IES Color Committee (2020). *ANSI/IES TM-30-20 Technical Memorandum: IES Method for Evaluating Light Source Color Rendition.* New York: Illuminating Engineering Society.

IES Color Committee (2021) *ANSI/IES LS-5-21 Lighting Science: Color.* New York: Illuminating Engineering Society.

IES Nomenclature Committee (2020). *ANSI/IES LS-2-20 Lighting Science: Concepts and Language of Lighting.* New York: Illuminating Engineering Society.

Illuminating Engineering Society (2018). PS-11-18 IES Position on TM-30-08, IES Method for Evaluating Light Source Color Rendition. Available at: https://www.ies.org/advocacy/position-statements/ps-11-18-ies-position-on-tm-30-18-ies-method-for-evaluating-light-source-color-rendition/ [accessed December 13, 2023]

International Commission on Illumination (1995). *CIE 13.3:1995 Method of Measuring and Specifying Colour Rendering Properties of Light Sources.* Vienna: International Commission on Illumination.

International Commission on Illumination (2014). *CIE TN 001:2014 Chromaticity Difference Specification for Light Sources.* Vienna: International Commission on Illumination.

International Commission on Illumination (2017). *CIE 224:2017 Color Fidelity Index for Accurate Scientific Use.* Vienna: International Commission on Illumination.

International Commission on Illumination (2018). *CIE 015:2018 Colorimetry, 4th Edition.* Vienna: International Commission on Illumination.

Livingston, J., Whitehead, L., and Royer M. (2021). All About Appearances: Avoiding Mismatches in White Light. *Lighting Design and Application* 51(1): 20-21.

MacAdam, David, L. (1942) Visual Sensitivities to Color Differences in Daylight. *Journal of the Optical Society of America*, 32:5, 247-274.

National Electrical Manufacturers Association (2017). *ANSI/NEMA C78.377 American National Standard for Electric Lamps – Specifications for the*

Chromaticity of Solid-State Lighting Products. Rosslyn: National Electrical Manufacturers Association.

Royer, M. (2022) Tutorial: Background and Guidance for Using the ANSI/IES TM-30 Method for Evaluating Light Source Color Rendition, *LEUKOS*, 18:2, 191-231, DOI: 10.1080/15502724.2020.1860771

4

Traditional Light Sources

Abstract

While traditional (i.e., non-LED) light sources are being phased out across the world, there are still millions in use. This chapter discusses the technologies that traditional sources, including incandescent, halogen, fluorescent, and metal halide, use to produce light. The chapter provides details about the characteristics, shapes, and bases typical of each technology, complete with illustrations of each.

Keywords: Efficacy, incandescent, halogen, fluorescent, compact fluorescent, linear fluorescent, HID, high-intensity discharge, metal halide, ceramic metal halide, high-pressure sodium, induction, lamp shapes, A-lamp, B-lamp, C-lamp, CA-lamp, G-lamp, T-lamp, MR-lamp, R-lamp, PAR-lamp, AR-lamp, beam angle, field angle, ballast.

4.1 Introduction

The rapid development of LEDs during the 2010s included dramatic improvements in their energy efficiency and color qualities. As a result, they quickly became the default light source for both indoor and outdoor lighting. Today, older traditional light source technologies, sometimes referred to as legacy technologies, are in steep decline due to two factors. First, a growing list of countries are eliminating older technologies by outright banning them or by setting energy efficiency requirements that they cannot meet, but that LEDs can. For example, the U.S. Department of Energy issued a ruling that, beginning in 2028, all general service lamps (which include A-lamps, directional lamps, specialty lamps, and some LED tube lamps) must have an efficacy of at least 83 lumens per Watt. This effectively eliminates all lamp technologies except LED (Appliance and Equipment Standards Rulemakings and Notices).

Second, as sales decline manufacturers are ending the production of various light source technologies and lamp shapes to focus on LEDs.

While these older technologies have little or no place in new lighting designs, there remain millions upon millions of these lamps installed across the world, and energy auditors encounter them every day when reviewing existing installations. This chapter will cover traditional light sources from that perspective, with a focus on understanding them as part of an assessment of existing installations to understand their properties in preparation for considering LED replacement options – either retrofit lamps or new luminaires.

4.2 Lamp Efficacy and Lamp Life

Two key measurements for all lamps are their efficacy and rated life. For traditional light sources, the rated lamp life is the length of time that has passed when one half of a sample set of lamps has failed. LED lamp life is discussed in Chapter 5. The environmental and measurement conditions are carefully controlled and are specified in standards published by National Electrical Manufacturer's Association (NEMA) and the Illuminating Engineering Society (IES) for each lamp technology (incandescent, fluorescent, etc.).

Efficacy is often confused with efficiency, but they are not the same thing. *Efficiency* is a unitless ratio comparing two like things. For instance, the efficiency of a light fixture, or luminaire, is the amount of light it emits divided by the amount of light created by the internal light source. A luminaire with a light source that generates 1000 lumens but that emits 900 lumens is 90% efficient (the other 100 lumens are absorbed by various internal components).

Efficacy is a ratio comparing two different things, and therefore has units. The efficacy of a light source is the amount of light generated, measured in lumens, for each unit of energy consumed, measured in watts. A standard incandescent lamp might produce 10 lumens of light for every watt of power consumed, so it would have an efficacy of 10 lm/W (lumens per Watt).

4.3 Incandescent Lamps

Incandescence is the emission of light by an object resulting from its high temperature. Light generated by a candle is an example of incandescence. Electric light sources produce light by passing electricity through a material (the tungsten filament) that has a high electrical resistance. The resistance causes the material to heat to the point of giving off light, or incandescing, hence the "incandescent" lamp.

4.3.1 Standard incandescent lamps

Early incandescent lamps of the 1880s used a bamboo filament to generate light. Over the years incandescent lamps saw incremental improvements, such as switching to tungsten for the filament, but even at their most efficacious (energy efficient) only about 10% of the energy consumed is turned into visible light resulting in an efficacy of 8–13 lumens per Watt (lm/W) (Livingston, 2022). In addition to having the lowest efficacy of any electric light source, standard incandescent lamps also have the shortest life. Table 4.1 provides a summary of light source efficacy and average life (IES Light Sources Committee, 2020, Livingston, 2022).

The construction of an incandescent lamp is fairly simple, as shown in Figure 4.1. A tungsten filament is held in place by two rigid wires that also conduct electricity to it. These lead/support wires are attached to a metal base, which fits into the luminaire's socket to support the lamp and provide an electrical connection. A glass envelope sits on top of the base and around the filament and is filled with inert gases, typically argon and nitrogen, to prevent the filament from igniting.

As noted earlier, these lamps are rapidly being taken out of production. In assessing an existing installation, it is best to assume that replacement incandescent lamps are not available but that LED retrofit lamps are. Even if a retrofit lamp is available, it is often more energy efficient to replace a general illumination fixture, such as a downlight (see Chapter 6) with an LED luminaire than to use a retrofit lamp. Planning such a replacement is best done with a knowledgeable and experienced lighting retrofit practitioner or designer, who may be able to simultaneously improve the quality of the

Table 4.1 Light sources, efficacy, and average life.

Light source technology	Efficacy (lm/W)	Average life (hrs)
Wax candle	0.3	NA
Incandescent	8–13	750–1500
Halogen	10–36	3000–5000
Compact fluorescent	35–80	6000–8000
Linear fluorescent	70–100	15,000–45,000
Metal halide	70–125	10,000–30,000
White light LED (as of early 2024)	60–140	30,000–70,000

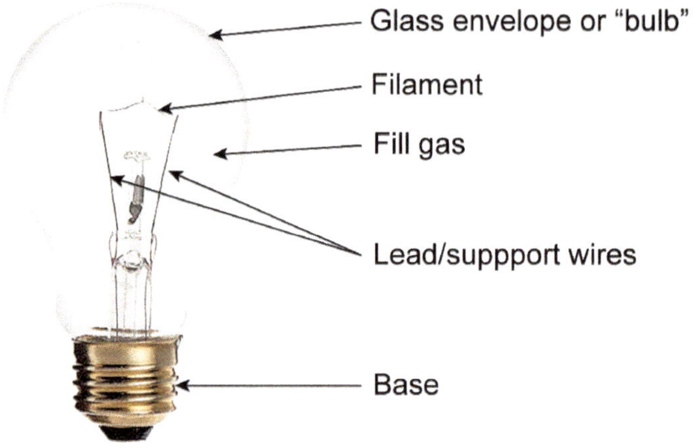

Figure 4.1 Parts of an incandescent lamp.
Photo courtesy of Satco Products, Inc.

overall lighting design rather than simply make a one-for-one replacement. Many decorative fixtures, on the other hand, are still manufactured with screw bases, so the use of LED retrofit lamps is expected.

4.3.2 Halogen lamps

The early 1960s saw the introduction of the halogen lamp (also variously marketed as tungsten-halogen, quartz-halogen, and quartz lamps), which differed from the older incandescent lamp in two ways. First, the fill gas is a halogen, typically iodine or bromine, under pressure. To resist the pressure the envelope is made of a harder, tempered quartz glass (the quartz part of quartz-halogen) (IES Light Sources Committee, 2020).

The halogen fill gas provides two benefits. The first is a higher efficacy than standard incandescent lamps at between 10 and 35 lm/W. The second is a longer life – often twice as long as a comparable incandescent lamp. Halogen lamps are smaller than incandescent lamps. Often, the halogen version of a traditional incandescent shape has a small halogen capsule inside the larger glass envelope, as shown in Figure 4.2.

An improved version of the halogen lamp is the halogen infrared (HIR) lamp. It uses a special coating inside the envelope to reflect infrared light back to the filament, increasing the temperature and therefore the efficacy without consuming more electricity.

Figure 4.2 Halogen capsule in an A-shaped envelope.
Courtesy of Philips, a global brand of Signify

Halogen lamps are treated the same way as incandescent lamps for the purposes of replacing fixtures and using retrofit LED lamps.

4.3.3 Incandescent and halogen lamp shapes and sizes

At the peak of their use, incandescent lamps were available in over a dozen shapes, and every shape was available in a range of sizes. Later many of those shapes were outfitted with halogen light sources. Today most traditional lamp shapes are still being manufactured, especially those used for decorative purposes, but use LEDs as the light source. The shapes can be separated into two general groups – omnidirectional lamps and directional lamps. All shapes are identified by one to three letters which are derived from a description of the shape. Depending on the shape, the glass envelope may be available as clear, frosted, and/or colored.

4.3.3.1 Omnidirectional lamp shapes

Omnidirectional lamps emit light in all directions. If there is a need to control the light or give it direction the control comes from the luminaire, or an accessory attached to it. The most common omnidirectional lamp shapes, shown in Figure 4.3, are:

- A (arbitrary) lamps are the traditional lamp shape that has existed since the 1870s.
- B (blunt tip or bullet) lamps are often exposed in decorative fixtures such as chandeliers and wall sconces.

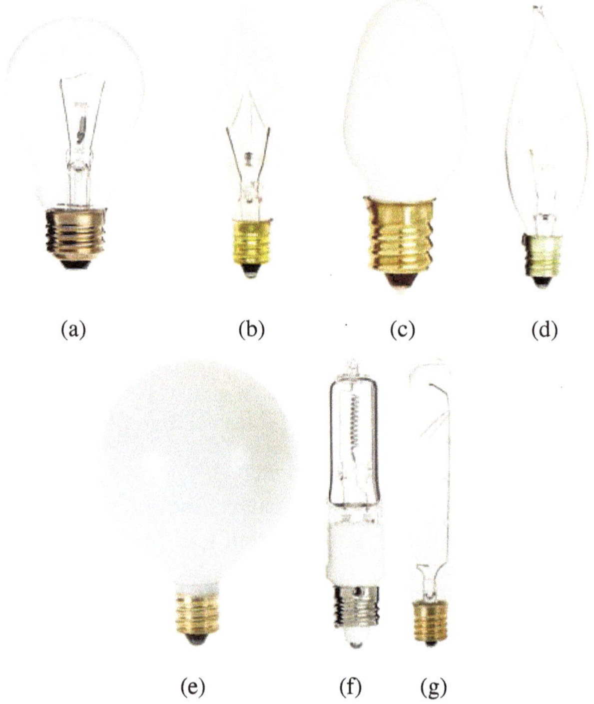

Figure 4.3 Omnidirectional lamp shapes. (a) A-lamp (b) B-lamp (c) C-lamp (d) CA-lamp (e) G-lamp (f and g) T-lamps (not to scale).
Courtesy of Satco Products, Inc.

- C (cone) lamps have typically been used in night lights and holiday decorations.
- CA (candle flame) lamps, like B-lamps, are often exposed in decorative fixtures such as chandeliers and wall sconces. The difference is that they have a glass flame tip at the top of the envelope.
- G (globe) lamps are also frequently used in decorative fixtures and in theatre marquees.
- T (tube) lamps are frequently used in picture lights, reading lights, and similar fixtures.

4.3.3.2 Directional lamp shapes

Traditional directional lamps include a reflector in their design and may include a lens. While the lamp provides most of the beam control, the luminaire may hold accessories that provide additional control. The exact

(a) (b) (c) (d)

Figure 4.4 Directional lamp shapes. (a) R lamp (b) PAR lamp (c) MR lamp (d) AR lamp (not to scale).

(a, c) Courtesy of Satco Products, Inc.
(b, d) Courtesy of Philips, a global brand of Signify

beam angle and associated description vary from one manufacturer to another. Typical beam angle descriptions and angle ranges are shown in Table 4.2. Directional lamp shapes, shown in Figure 4.4, include:

- R (reflector) lamps are the original incandescent directional lamp shape. Historically, they have been available in two beam angles – spot and flood.
- PAR (parabolic aluminized reflector) lamps place the filament at the focal point of the parabolic reflector. This improves control of the reflected light and permits the efficient use of lenses to create a wide range of beam angles.
- MR (multi-reflector) lamps have a bowl-shaped reflector with many facets (each considered to be an individual reflector) that are shaped and

Table 4.2 Beam angle descriptions and designations.

Description	**Designation**	**Beam angle (varies by manufacturer)**
Very narrow spot	VNSP	<10°
Narrow spot	NSP	5–15°
Spot	SP	5–15°
Narrow flood	NFL	20–30°
Flood	FL	25–40°
Medium flood	MFL	30–45°
Wide flood	WFL	40–50°
Very wide flood	VWFL	50–60°

angled to reflect light to produce the desired beam angle. These lamps are typically low voltage (12 V) and are small, with the most common size only 2” (50 mm) in diameter.

- AR (aluminum reflector) lamps are unique in two ways. First, the reflector is stamped out of polished aluminum, not shaped glass with a reflective coating. Second, the light source is partially covered so that the only light leaving the lamp bounces off the reflector. This decreases efficacy but increases beam control. The reflector is typically faceted, like that of an MR-lamp, to control the beam angle.

4.3.3.2.1 Beam and field angles

Directional lamps produce a cone of light. In all cases, the center is the brightest part of the beam and brightness declines toward the edge of the beam. Therefore, the brightness at the center of the beam (regardless of the actual brightness measurement) is described as 100%. The beam angle is the angular measurement taken across the beam at the point where the brightness is 50% of center beam brightness, as shown in Figure 4.5. When manufacturers list the beam angle of a directional lamp or a luminaire, that is what it means. Note that this is not necessarily 50% of the width of the cone of light. In fact, beam angle and the angle of the full cone of light have no relationship to one another.

There is light outside of the beam angle. The measurement across the beam at the point where the light has fallen to 10% of center beam brightness is called the field angle. Lamp and fixture manufacturers do not always provide the field angle measurement. As with beam angle, the field angle and the full cone of light have no relationship to one another.

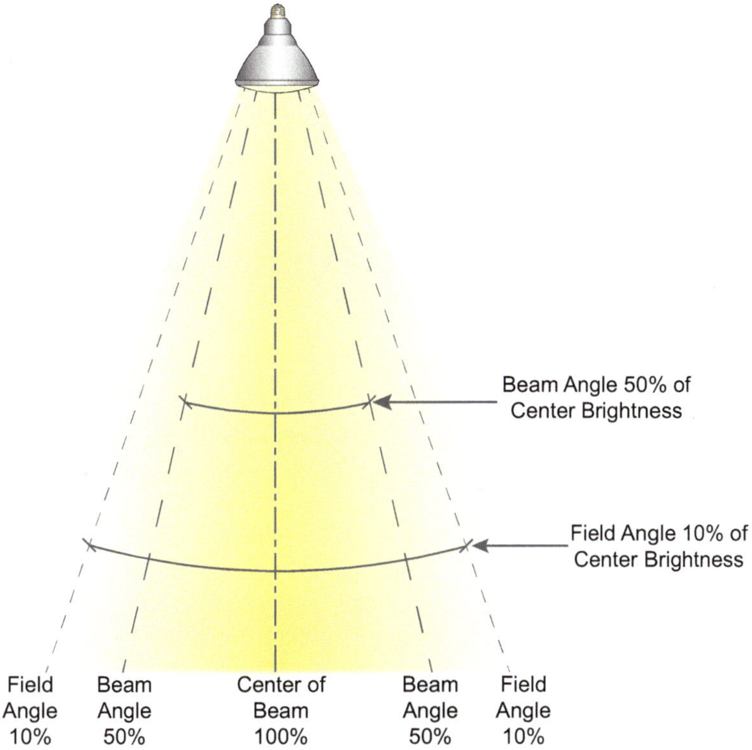

Figure 4.5 Beam and field angles.

Light beyond the field angle has no name because it is considered insignificant. It will seamlessly blend into the beams of adjacent fixtures and contributes very little to the brightness of a space.

4.3.3.3 Describing lamp shapes and sizes

Most shapes are available in several sizes, which are determined by the measurement of their largest diameter. The lamp is described by the shape followed immediately by the size. In the U.S. the unit of measurement is 1/8". A typical A-lamp has a diameter of 2 3/8" so it would be described as A19. In most of the rest of the world the unit of measurement is 1 mm, so a similar lamp (with a diameter of 60 mm) would be described as A60.

A more complete lamp description adds the wattage at the beginning. A 50 W version of the lamps above would be described as 50A19 and 50A60. Some manufacturers of LED retrofit lamps continue to use this nomenclature, while others deviate from it or include additional descriptors. For example, one manufacture's description of an 8.5 W LED replacement for a 75 W incandescent A-lamp is 8.5A19/LED/930/FR/P/ND 4/1FB where 8.5 is the wattage, A19 is the size, and LED indicates an LED retrofit lamp; 930 indicates the color temperature and color rendering (Chapter 3) where 9 indicates a CRI in the 90s and 30 indicates a nominal CCT of 3000 K.

All directional lamp shapes are available with two or more beam angles, so a description of the cone of light must be part of the lamp designation. The beam angle description is added to the shape/size description. A 50 W PAR30 (parabolic aluminum reflector with a diameter of 3 6/8") with a 25° beam would be described as a 50PAR30FL. Sometimes the beam angle is separated by a slash. Since the "FL" designation is not very specific, many manufacturers add the actual beam angle after the description. In that case, the lamp just described would be a 50PAR30/FL25.

4.3.3.4 Incandescent lamp bases and sockets

Figure 4.1 shows that the base is the part of the lamp where it connects to the luminaire and power. Many incandescent lamp shapes are fitted to more than one base, so it becomes necessary to know the various bases and corresponding sockets. Figure 4.6 illustrates the bases that may be attached to an incandescent or halogen lamp shape.

4.4 Fluorescent Lamps

Fluorescence occurs when certain materials absorb the energy of short wavelength, high-energy light (typically ultraviolet or UV) and then emit a portion of that energy at longer wavelengths and lower energy. In fluorescent lamps an arc of electricity is established between electrodes at either end of the lamp. The electricity is carried on a fill gas, such as argon or krypton, which is saturated with vaporized mercury. When electrified, the mercury gas emits UV light. To convert the UV light to visible light the inner walls of the fluorescent tube are lined with phosphors, either naturally occurring or man-made. Three phosphors are typically used, one producing light in the blue range, one in the green range, and one in the orange-red range. The mix and balance of phosphors determines the color temperature and color rendering of the resulting light. (IES Light Sources Committee, 2020, Livingston, 2022)

Line voltage (the voltage supplied to the fixtures) isn't sufficient to create and sustain the arc inside the fluorescent tube. To do so, all fluorescent lamps require a supplemental device called a ballast. The ballast modifies the

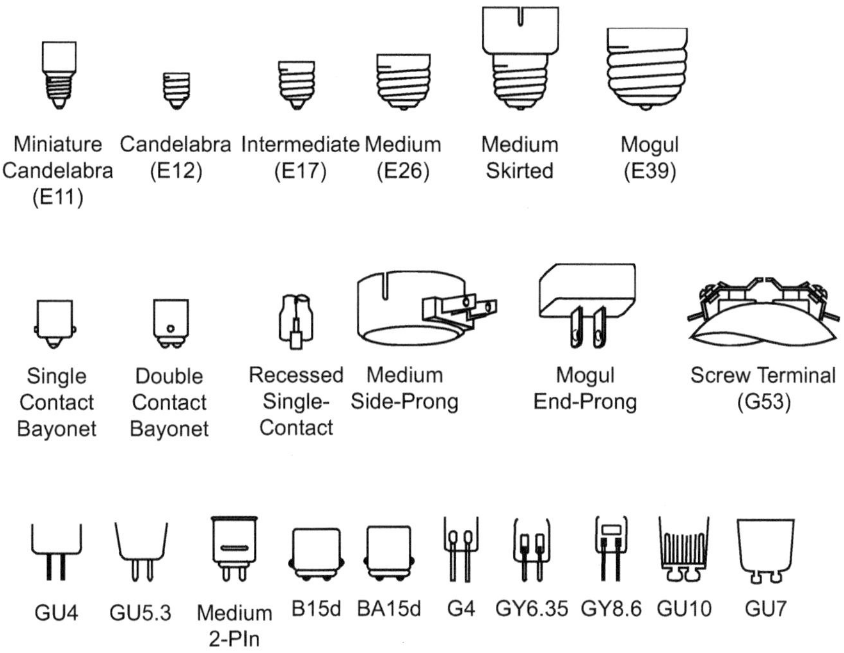

Figure 4.6 Incandescent and halogen lamp bases (not to scale).
Courtesy of Philips, a global brand of Signify

electricity to create and sustain the arc and regulates the electricity flowing through the fluorescent tube while in use. Ballasts have very long lives, much longer than the life of the lamps, so they are installed as a part of the luminaire, not part of the lamp.

Although fluorescence is a more complicated process than incandescence, it is also more energy efficient. At the peak of their use, fluorescent lamps had an efficacy range of 55–100 lm/W with lifetimes of up to 45,000 hours. Compared to the efficacy and life of incandescent lamps, it's easy to see why, by the 1950s, so many fluorescent lamps were in use that they were creating more total light than incandescent lamps (History of the Light Bulb).

In November of 2023 the Minamata Convention on Mercury, a program of the United Nations with delegates from at least 150 countries, agreed to phase out fluorescent lighting globally and completely by 2027 (The World Agrees On Full Phase-Out, 2023). Some countries and/or manufacturers may not meet the deadline. However, lighting practitioners should assume that fluorescent lamps are no longer available and search for LED retrofit lamps when relamping existing luminaires is part of a project.

4.4.1 Linear fluorescent lamps

Fluorescent lamps were originally available only as straight tubes of glass, called linear fluorescents. Until the 1980s these tubes were almost exclusively 1 $\frac{1}{2}$" (40 mm) in diameter. These lamps are described by the letter "T" for tube followed by the diameter measured in $\frac{1}{8}$" or 1 mm increments, so they were called T12 or T40 lamps. The 1980s saw the introduction of smaller and more efficient T8 lamps, followed by T5 lamps in the 1990s. While other sizes were available, these dominated the market.

Two variations on the linear fluorescent were developed. In the first, called a u-bend lamp, the tube is bent into a "U" shape, with the legs of the U either 4" or 6" on center. In the second, the tube is bent into a full circle and is called a circle-line. See Figure 4.7 for fluorescent lamp shapes and Figure 4.8 for bases.

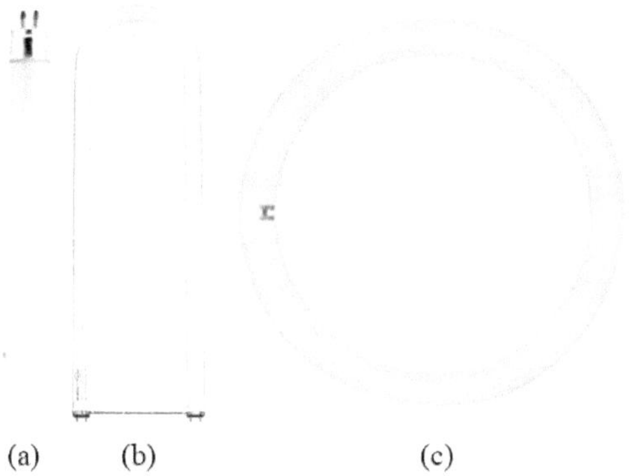

Figure 4.7 Linear fluorescent lamp shapes. (a) linear (b) u-bend (c) circle-line (not to scale). Courtesy of Satco Products, Inc.

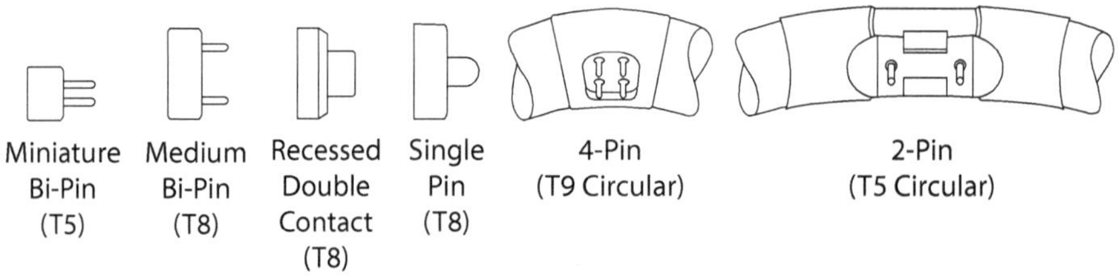

Figure 4.8 Linear fluorescent lamp bases (not to scale). Courtesy of Philips, a global brand of Signify

4.4.2 Compact fluorescent lamps

In the 1980s smaller fluorescent lamps were introduced, some small enough to replace traditional A-lamps. These were called compact fluorescent lamps or CFLs, and were available in a wide range of shapes, sizes, and wattages. As manufacturers introduced more and more fixtures that could use these lamps, they became the default light source for small aperture downlights in commercial lighting projects.

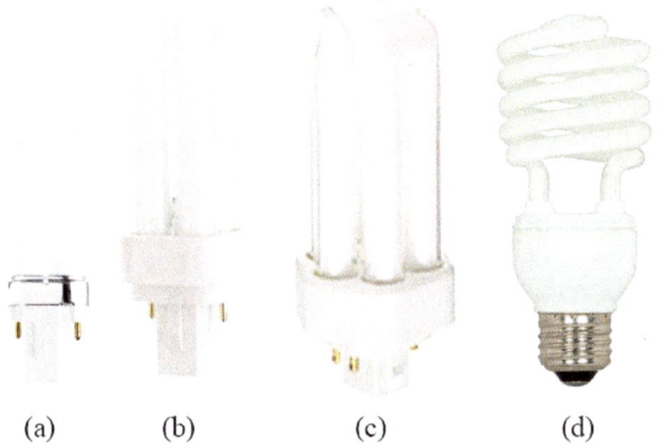

Figure 4.9 (a) Single U (PL or Biax) (b) double U (PL-C or Double Biax) (c) triple U (PL-T or Triple Biax) (d) twist or swirl retrofit (not to scale).
Courtesy of Satco Products, Inc.

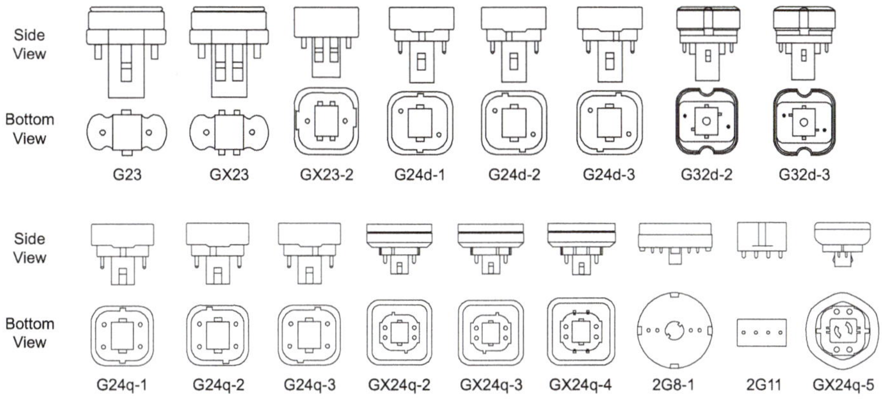

Figure 4.10 Compact fluorescent lamp bases (not to scale).
Courtesy of Philips, a global brand of Signify

While the shapes of CFL lamps are consistent across manufacturers and the lamps are interchangeable, the nomenclature of CFLs was never standardized across the industry. Instead, each manufacturer used their own labeling system (Philips called theirs PL lamps, GE called theirs Biax, etc.). The basic shape is a long narrow U. To limit length but increase output, manufacturers would place two or three U shapes on a single base, as shown in Figure 4.9. Additionally, the most common shape for retrofit CFL lamps is the "twist" or "swirl." LED retrofit lamps for all three U-shaped CFL lamps are available. The replacement for the swirl lamp is an LED A-lamp. Pin bases for compact fluorescent lamps without integral ballasts are shown in Figure 4.10.

4.4.3 Ballasts

Ballasts are necessary for the operation of fluorescent and high-intensity discharge lamps. Ballasts provide the initial voltage to create an arc of electricity between the lamp's electrodes, and regulate the flow of electricity to maintain the arc. Early fluorescent ballasts were magnetic. They used a coil of wire around an iron core. The technology is simple, but not very energy efficient. In the U.S. their phase out began around 2010 when they could no longer be used in new fixtures. Today, a few manufacturers still produce magnetic ballasts that can be used for repairs, but they are increasingly difficult to find.

Later ballasts performed the same function but used electronics, hence the name electronic ballast. Electronic ballasts are approximately 10% more efficient than magnetic ballasts (IES Light Sources Committee, 2020).

Most fluorescent lamps can be paired with a dimming ballast. The dimming ballast must be paired with the correct control technology (reverse phase dimming, 0–10 V control, etc., see Chapter 9) to operate properly (Livingston, 2022).

4.5 High-intensity discharge

High-intensity discharge (HID) lamps are similar to fluorescents in that they pass an electric arc through a gas-filled enclosure. However, rather than create ultraviolet and convert it to visible light, as fluorescents do, the gases and other materials in an HID lamp produce light directly. This process is more energy efficient than fluorescence. As with other traditional light technologies, while HID lamps are still being produced the assumption should be that LED retrofit lamps are a better long-term choice.

4.5.1 Metal halide

Metal halide lamps use combination of argon, neon, and/or xenon as the fill gas. The enclosure also contains mercury and metals or metal halides that, when electrified, produce visible light. Two types of metal halide lamps are available and are defined by the material enclosing the gases and metals. The enclosure was originally made of quartz, and lamps using quartz are called quartz metal halide or MH lamps. The other material is clear ceramic used in ceramic metal halide or CMH lamps. The light-producing capsule is held within a hard glass outer enclosure, as shown in Figure 4.11.

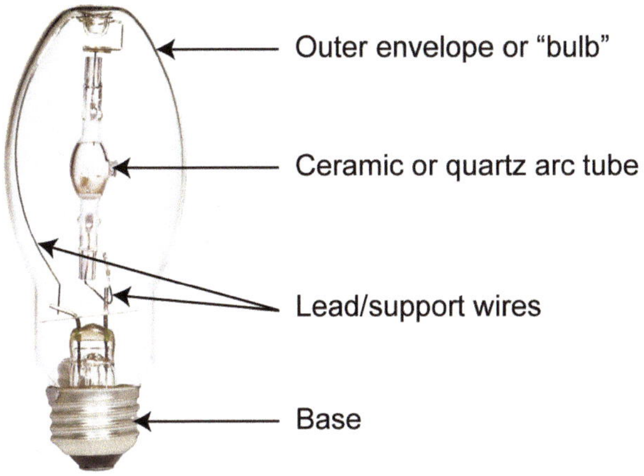

Figure 4.11 Parts of a metal halide lamp.
Photo courtesy of Satco Products, Inc.

4.5.2 High-pressure sodium

High-pressure sodium (HPS) lamps use a ceramic arc tube containing a sodium-mercury mixture that partially vaporizes during operation. The arc tube is enclosed in a hard glass envelope. HPS lamps are known for their warm light (approximately 2200 K) and poor color rendering (IES Light Sources Committee, 2020). These drawbacks were offset by their high efficacy and long life, which is why they were often used to light roadways and parking lots. As with other discharge lamps, they require a ballast. LED replacements for high-pressure sodium lamps are available.

4.5.3 Low-pressure sodium

Low-pressure sodium (LPS) lamps use an arc tube containing sodium that is enclosed in a hard glass envelope. While they are energy efficient, they produce a monochromatic yellow light that is roughly equal to 1800 K. The extremely poor color rendering is not balanced by the high efficacy, so these lamps have been replaced by other technologies in all applications (IES Light Sources Committee, 2020).

4.5.4 HID lamp shapes

The relatively small capsule in HID lamps can be placed inside many traditional lamp shapes, including A, MR, R, PAR, and T. MH and CMH lamps in these shapes were available for low wattage interior fixtures such as downlights and track lights. Larger shapes for higher wattage fixtures, such as streetlights, were available in shapes called ED, BD, and BT, as shown in Figure 4.12. HID lamp bases are shown in Figure 4.13.

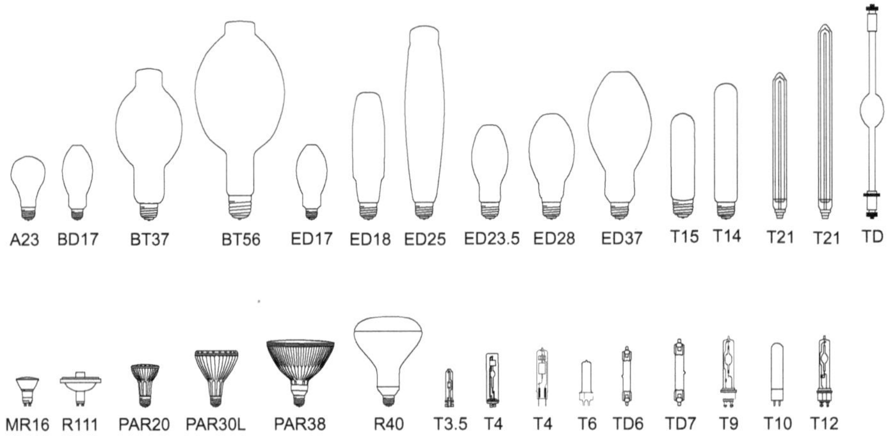

Figure 4.12 HID lamp shapes (not to scale).
Courtesy of Philips, a global brand of Signify

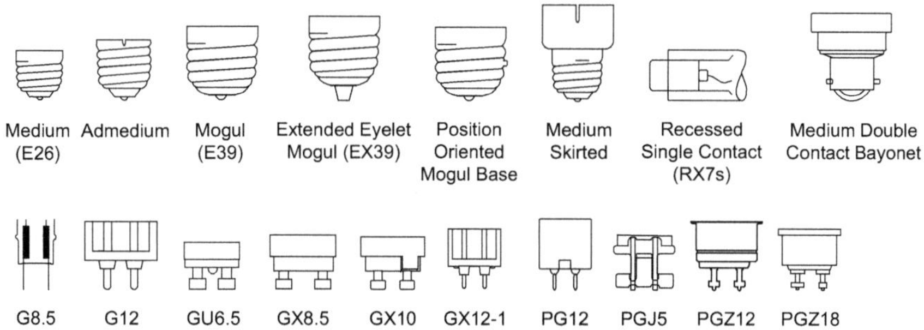

Figure 4.13 HID lamp bases (not to scale).
Courtesy of Philips, a global brand of Signify

4.6 Induction

Induction lamps are a variation on fluorescent lamps and have very similar properties. The main difference is that they do not have electrodes or ballasts. Instead, an induction lamp generates light by exciting the mercury inside the lamp with a radio frequency or microwave generator. In fluorescent lamps the electrodes are a likely point of failure. Without them, induction lamps can have lifetimes of up to 100,000 hours, making them an excellent choice for hard-to-reach locations. Unlike fluorescent lamps, they also work well at low temperatures. However, due to induction lamp's size, significant lamp lumen depreciation, and possibly shorter lifetime, LEDs are now a better choice.

References

Appliance and Equipment Standards Rulemakings and Notices (n.d.). Available at: https://www1.eere.energy.gov/buildings/appliance_standards/standards.aspx?productid=4. [accessed April 22, 2024].

History of the Light Bulb. Available at: https://www.energy.gov/articles/history-light-bulb (accessed October 12, 2023).

IES Light Sources Committee (2020) *ANSI/IES LP-4-20 Lighting Practice: Electric Light Sources -Properties, Selection, and Specification.* New York: Illuminating Engineering Society.

Livingston, Jason. (2022). *Designing with Light: The Art, Science, and Practice of Architectural Lighting, 2nd Ed.* Hoboken: John Wiley & Sons, Inc.

The World Agrees On Full Phase-Out Of Fluorescent Lamps But Fail To Catch Full Savings (November 7, 2023) Available at: https://www.eceee.org/all-news/news/the-world-agrees-on-full-phase-out-of-fluorescent-lamps-but-fail-to-catch-full-savings/ (accessed December 4, 2023)

5

LED Light Sources

Abstract

LEDs are the dominant light source of today and the foreseeable future. This chapter introduces both light-emitting diodes (LEDs), and organic light-emitting diodes (OLEDs) and the technologies behind how they produce light. It explains the parts of an LED package and the additional components needed for an LED to operate, and discusses environmental factors related to using LEDs.

Keywords: Light emitting diode, LED, organic light emitting diode, OLED, driver, binning, phosphor, COB, LM-80, TM-21, n-layer, p-layer, narrow band emitter, blue pump LED, pc-LED, surface-mounted device, L70, solid state lighting.

5.1 Introduction

The first practical light emitting diode (LED), which produced red light, was invented by Nick Holonyak at General Electric in 1962. Over the next 30 years other colors of LEDs were developed and were used as indicator lights, in digital clocks, etc. While useful for other purposes, none of them were bright enough to be used as a source of illumination. In 1993 Shuji Nakamura, Isamu Akasaki, and Hiroshi Amano developed the high-brightness blue LED, for which they were awarded the Nobel Prize in Physics in 2014. The blue light LED became the foundation for the white light LEDs in world-wide use today.

In less than a decade LEDs went from being an interesting, if poor quality, source of light to the dominant light source in new luminaires and replacement lamps. Their efficacy exceeds that of all other light sources and continues to increase. For example, the U.S. Department of Energy has set a goal of white light luminaires with efficacies of 250 to 275 lumens per Watt (LED Basics), which is more than 20 times as efficient as incandescent lamps and more than double the most efficient fluorescent lamps.

5.2 Basics of LED Construction and Operation

Unlike other electric light sources, LEDs do not need or use a gas-filled chamber to produce light. They are solid, semiconductor devices (which is why they are sometimes referred to as solid state lighting) that produce ultraviolet, visible, or infrared light, depending on the materials used, when electricity passes through them.

LEDs are manufactured by coating a base or substrate layer with various materials to create a large semiconductor. The semiconductor is then cut into very small pieces, which are the individual LEDs. The fundamental LED consists of two layers on top of the base layer. The first layer is called the n-layer because it is negatively charged. The second layer is called the p-layer because it is positively charged. Where they touch, electrons cross over from the n-layer to the p-layer to create a neutral zone. When direct current is supplied to the n-layer it forces electrons into the p-layer where they temporarily fall into electron "holes" surrounding the atoms and then are ejected. When this happens, energy is released in the form of photons. The specific amount of energy released is determined by the materials used to make the LED, but in all cases the range is relatively narrow. The result is that each LED produces light of a narrow band of wavelengths, which is why LEDs are sometimes called narrow-band emitters. Various combinations of materials can be used, but in all combinations a narrow band of colored light is produced. No combination of materials in an LED produces white light.

The most common method of producing white light uses a blue LED, (often referred to as the "blue pump," although UV and indigo LEDs can also be used) topped with a broadband phosphor, similar to the phosphors used in fluorescent lamps. The phosphor is stimulated by the blue LED and converts much of the blue light to light of longer wavelengths that span the visible spectrum (Figure 5.1).

The resulting SPD is perceived as white light. Phosphor-converted LEDs (pc-LEDs) are recognizable due to the distinct spike in the blue end of their SPD, and a broad hump from cyan through red, as shown in Figure 5.2.

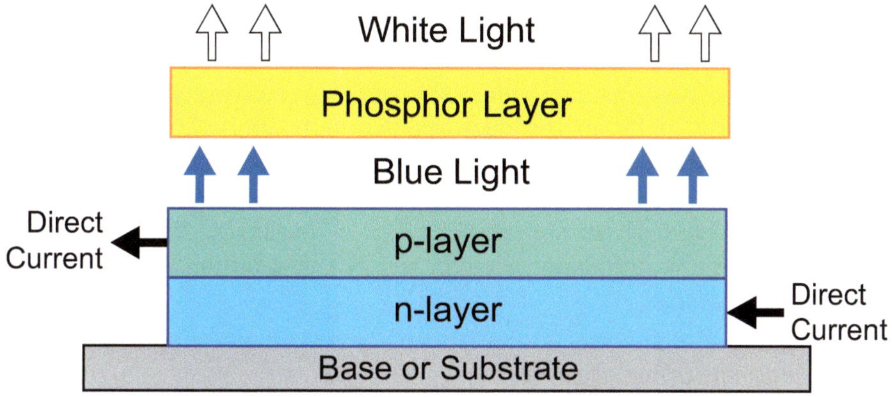

Figure 5.1 Schematic diagram of a white light LED.

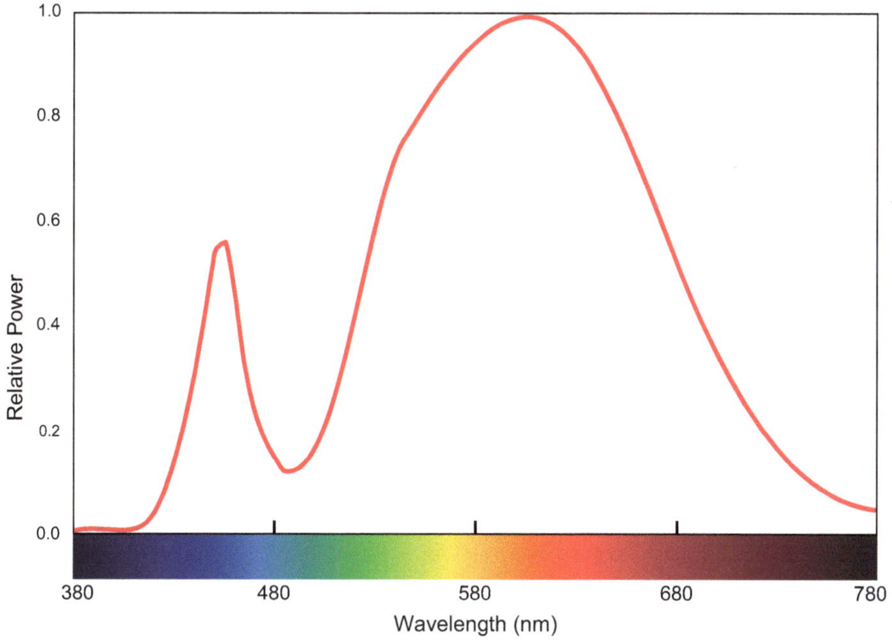

Figure 5.2 SPD of a pc-LED showing the characteristic spike of energy from the blue LED and a broad distribution of wavelengths from the phosphor.

5.3 LED Packages

The LED is sometimes referred to as a "die" or "chip." LEDs are sensitive to heat, which can lead to color shifts, reduced light output, and shortened life. To combat these effects, LED dies are mounted on a heat sink to draw away the excess heat. LEDs also produce a cone of light, around 120°, which is too broad for most applications, so they are usually fitted with a lens. The combination of LED, heat sink, lens, and other components is often described as an LED "package." There are several types of packages. The type of package chosen by the luminaire manufacturer is based on the luminaire's purpose or requirements.

5.3.1 Surface-mounted device

Surface-mounted device (SMD) LEDs are an improvement over the original method of mounting LEDs that had wire leads. SMD LEDs have small metal contacts that permit them to be soldered to circuit boards, which can lead to smaller products, faster manufacturing, and longer life (Introduction of SMD LED, 2022). LED tape and sheets are examples of SMD use.

5.3.2 Chip-on-board

Chip-on-board (COB) products combine multiple LED chips on a circuit board to form a single LED module. The array of LED chips is treated as a single, larger LED which allows them to be tightly spaced and produce more lumens in a smaller area. The advantages of COB packages include fewer components, a lower failure rate, and decreased light loss (The Basics of Chip on Board COB LEDs, 2016).

5.4 Organic Light-emitting Diodes

Organic LEDs (OLEDs) can be manufactured from a variety of organic (carbon-based) electroluminescent materials. They are like chip LEDs in that they have positive and negative layers and light is produced when electrons are forced through a light-emitting material. There are two unique aspects to OLEDS. The first is that they can be manufactured as a thin, lightweight, luminous surface – a planar light source rather than a point source-that can provide a low-glare luminous surface. The second is that they can be produced with a flexible substrate which allows them to be curved or twisted to produce luminaire shapes that are not possible with other light sources (Livingston, 2022).

5.5 Binning

The process used to coat the LED substrate with electroluminescent materials inherently creates variations in performance. To limit the amount of waste, LED manufacturers test and sort each LED die for several characteristics, including light output, color, and operating voltage. LEDs with similar characteristics are grouped into "bins." Binning by LED manufacturers lets luminaire manufacturers be more precise in the LEDs they purchase by selecting a set of characteristics that meet the requirements for their luminaire or product line (White Paper: Binning and LED).

As noted in Chapter 3, the most common way of communicating the LED variation a luminaire manufacturer will accept is with MacAdam ellipses. A manufacturer who uses a large ellipse, bins within a four-step ellipse for example, accepts a considerable amount of variation. Doing so likely leads to greater availability and reduced price for LEDs. It also likely leads to increased luminaire to luminaire variability, which may be considered undesirable, or even rejected, by the specifier or occupants.

5.6 Drivers

LEDs operate on direct current at reduced voltage. Drivers are the devices that convert incoming alternating current (AC) line voltage to direct current (DC), step down the voltage (often to 12 V or 24 V), and regulate the outgoing current and voltage. In most luminaires the driver is built in, but remote drivers are used for luminaires such as LED tape, and for luminaries in hard to access locations.

Most energy conservation codes require luminaires to have multiple output levels or to be dimmable, so most luminaire manufacturers provide dimmable drivers as the default option. This requires the lighting practitioner to ask and answer two questions.

The first question is the desired or acceptable minimum light level. Dimming drivers' descriptions include the minimum stable light output when dimmed. As of early 2024, 5% and 1% are common, with 0.1% frequently available. Five percent dimming means that the driver can sustain steady light output at 5% of maximum. Below 5% the driver's performance is unpredictable. The LED may flicker or may drop to 0%. Practitioners need to understand their client's needs and expectations and select a driver with the appropriate minimum dimming level.

It's also important to keep in mind that the human vision system is constantly adjusting to changes in illumination. Chapter 2 describes how, as the light in a room dims, the iris of the eye opens and the retina becomes more sensitive. The result is that dimming a room to 1% of maximum brightness when measured with a light meter will be perceived as 10% by the room's occupants.

The second question relates to the control system. There are several protocols used in architectural lighting control systems, including 0–10 V, DMX, and DALI, but most drivers accept only one. A mismatch between the driver and the control system results in a luminaire that cannot be controlled and could result in a costly remedy. Control protocols are discussed in Chapter 9.

5.7 LED Testing and Standards

LEDs differ from traditional light sources in several ways. One of them is that LEDs don't experience catastrophic failure or "burn out." Instead, LEDs dim as they age. At some point practitioners and occupants would say that the light provided has declined to such an extent that the luminaires are no longer useful and need to be replaced. The current standard for useful life of an LED is the time it takes (measured in hours of operation) for output to drop to 70% of initial output, a measure known as L70. Some applications won't tolerate a 30% reduction in illumination, so some manufacturers report lifetime to 80% of initial output (L80) and/or lifetime to 90% of initial output (L90).

LED packages (but not LED luminaires or LED retrofit lamps) are tested according to *ANSI/IES LM-80 Approved Method: Measuring Maintenance of Light Output Characteristics of Solid-State Light Sources*. LM-80 describes the testing methods and conditions for measuring the light output for LED packages, arrays, and modules over a time period. The test duration for LM-80 should be "consistent with the design operation life" of the LED chip and is typically 6000 hours or more. Notably, LM-80 does not provide a method of extrapolating the data to predict the life of an LED, and it is not used to measure LED luminaires.

At the end of the LM-80 test period an LED package may not have reached L70. *ANSI/IES TM-21 Technical Memorandum: Projecting Long-Term Luminous, Photon, and Radiant Flux Maintenance of LED Light Sources* defines the method of extrapolating the data collected in LM-80 testing to predict the useful life of an LED. The extrapolation method is limited to predicting lumen maintenance down to 70%, and to 6× the TM-80 testing period. This is important because some luminaire manufacturers make lifetime claims that can't be supported by the LM-80 data and TM-21 prediction. For example, a manufacturer who claims an L70 life of 100,000 hours would have needed to collect LM-80 data for at least 16,667 hours or 1.9 years. With the rapid ongoing development of LEDs, it's unlikely a manufacturer invested two years in LM-80 testing before releasing the LED product to the market. As of early 2024, L70 predictions of 30,000 to 50,000 hours are believable. L70 predictions of 100,000 hours probably are not. However, as LED production becomes more sophisticated and products more stable and durable, the industry may see the TM-21 6× limit raised to 10× or more. Lighting practitioners can consult TM-21 and industry publications for updates.

5.8 Environmental Factors

As described in Chapter 4, LEDs have a higher efficacy than traditional lighting technologies, meaning they produce more light while consuming less electricity, giving them a lower carbon footprint. LEDs' long life adds to their reduced operating costs because owners don't have the expense of replacement lamps and the associated labor costs. Combined, these two factors result in a lower life-cycle cost for LED luminaires (IES Light Sources Committee, 2020).

LEDs do not contain mercury or hazardous gasses. As a result, many LED luminaires are RoHS compliant. RoHS (Restriction of Hazardous Substances) originated in the European Union in 2006 and restricted the use of six hazardous materials found in electronic products. As of 2017, the list was expanded to cover 10 materials (RoHS Guide, 2024). RoHS is discussed in more detail in Chapter 10.

Unfortunately, it's not clear that LED luminaires reduce materials entering the waste stream or landfills. The housing of a well-made traditional luminaire could last 20 years or more. The fluorescent lamps and ballasts would be periodically replaced, for example, but the housing would last as long as the installation. The drivers of LED luminaires are replaceable, but the LED packages typically are not. When an LED luminaire reaches L70 or some other criteria indicating its light output is no longer adequate, and if the LED package is not replaceable, the entire luminaire must be replaced. As of early 2024, some manufactures have begun making commitments to provide replacement LED packages for a specified period of time – often five years – but that's much shorter than the potential life of the housing. The Zhaga Consortium is an organization seeking to standardize interfaces and connections for LED luminaire components (Zhaga Consortium). As of early 2024, nearly 100 manufacturers were listed in their database but few, if any, of them mention Zhaga Consortium participation in their marketing to North American practitioners.

A related problem is one of recycling. The manufacturers of many construction- related materials, such as those for ceiling tiles and carpet, have "take back" programs where they will take excess or used materials and recycle them into new products. Unfortunately, as of early 2024, no luminaire manufacturer has such a program. Aluminum and steel luminaire housings can be recycled, but it becomes the responsibility of the lighting practitioner, owner, or demolition contractor to identify the material and a recycler because the manufacturers offer no assistance. This obviously increases the likelihood that old luminaires are thrown away, not recycled.

References

The Basics of Chip on Board (COB) LEDs (2016). Available at: https://www.digikey.com/en/articles/the-basics-of-chip-on-board-cob-leds. [accessed on March 2, 2024].

IES Light Sources Committee (2020) *ANSI/IES LP-4-20 Lighting Practice: Electric Light Sources -Properties, Selection, and Specification.* New York: Illuminating Engineering Society.

Introduction of SMD LED: What is SMD LED? (2022). Available at: https://www.oasistek.com/en/News/what_is_smd_led#:~:text=A%20SMD%20LED%20is%20known,with%20epoxy%20resin%20or%20silicone..[accessedMarch2,2024]

LED Basics (n.d.). Available at: https://www.energy.gov/eere/ssl/led-basics. [accessed March 2, 2024].

Livingston, Jason. (2022). *Designing with Light: The Art, Science, and Practice of Architectural Lighting,* 2nd *Ed.* Hoboken: John Wiley & Sons, Inc.

RoHS Guide (2024). Available at: https://www.rohsguide.com. [accessed March 2, 2024].

White Paper: Binning and LED (n.d.). Holophane. (Note to Editor. This is an undated white paper on Holophane's site. They are a manufacturer, and a brand of Acuity Brands www.acuitybrands.com. I'm not sure how to show this document here.)

Zhaga Consortium (n.d.). Available at: https://www.zhagastandard.org. [accessed March 2, 2024].

6

Luminaires or Lighting Fixtures

Abstract

The most important piece of lighting hardware is the light fixture or luminaire. This chapter reviews the components of a luminaire and their function, then goes on to discuss the most common types of luminaires. It describes luminaire mounting categories and the several systems of categorizing luminaires based on how they distribute light in a space. Photos and line drawings illustrate each luminaire and distribution category.

Keywords: Luminaire, housing, reflector, aperture, flange, lens, diffuser, baffle, louver, blade, downlight, adjustable, accent, wall washer, wall grazer, cove light, track, track light, track head, recessed, semi-recessed, surface mount, pendant, direct, semi-direct, semi-indirect, indirect, diffuse.

6.1 Introduction

Once the outlines of the lighting design are understood – the application, target brightness, distribution of light across the space, aesthetic goals, budget, time frame, etc. – the lighting practitioner knows enough to begin selecting luminaires and testing layouts to see if they achieve the project's goals.

The number of manufacturers and lighting products available can seem overwhelming. For most practitioners, though, it is not necessary to know every manufacturer. Deep knowledge of a handful of manufacturers is usually a better approach, and allows the practitioner to become familiar with the manufacturer's performance, quality, cost, options and typical delivery times.

6.2 Classifying Luminaires

A luminaire is a complete light fixture that includes the light source. While it may seem as though the industry should have an agreed upon classification system to simplify communication, the truth is that there isn't one, although several organizations have systems for classifying certain types of luminaires. Instead, most manufacturers categorize their product line based on what they produce and how they market it (Livingston, 2022). In visiting a manufacturer's website, practitioners may find luminaires organized by:

- General luminaire type (downlight, wall washer, etc.)
- Application (residential, office, warehouse, etc.)
- Mounting (recessed, surface, pendant, wall, ceiling, etc.)
- Aperture shape and size (square, round, etc.)
- Light distribution pattern (see Section 6.7 later in this chapter)

6.3 Components of New Luminaires

Most luminaires are built with the same components, although the size, shape, and arrangement of those components can vary. With an understanding of a luminaire's components, lighting practitioners immediately understand how a luminaire functions, how it controls light, and how it might be modified if it isn't exactly what a project requires. These components are illustrated in Figure 6.1.

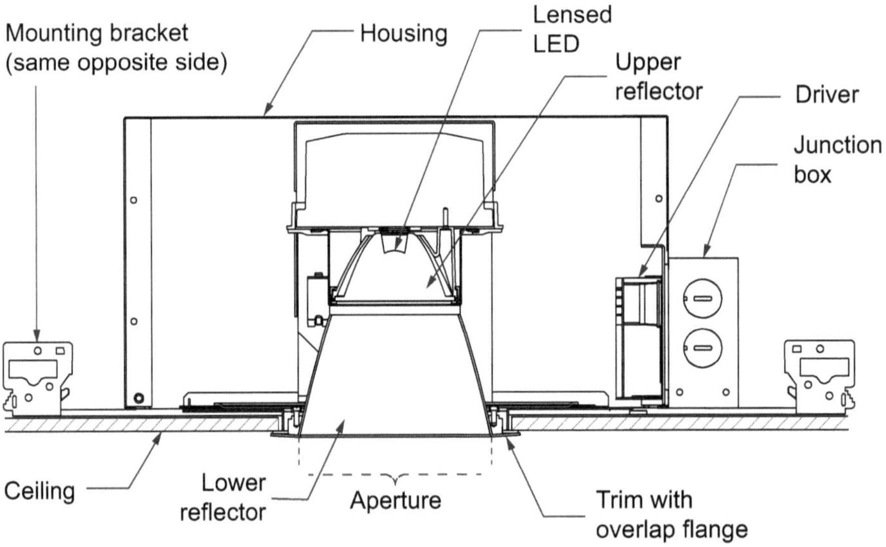

Figure 6.1 Common luminaire components.
Drawing courtesy of Acuity Brands Lighting, Inc.

6.3.1 Housing

A luminaire's housing serves several purposes. Historically, it was a metal box that effectively dissipated the heat of the incandescent lamp inside to prevent fires. Although this concern lessened with fluorescent and LED lamps, luminaire testing protocols still set maximum temperatures for the luminaire surface. Some form of housing, therefore, remains a requirement.

The second purpose of the housing is a mounting platform for the other components. The parts of a luminaire – the lamp holders, reflectors, lenses, drivers, etc. – all need to be securely attached to one another and/or to a frame that holds them in place. The housing also has mounting brackets where the housing is attached to the ceiling via several types of clips, and a junction box for connection to the building's electrical system.

6.3.2 Reflectors, flanged or flangeless

A reflector redirects light that would otherwise be lost inside the fixture housing so that it exits the luminaire through the aperture or strikes a lens or diffuser. In large flat luminaires such as a 2×2 (600×600) downlight, the reflector may be nothing more than a flat sheet of metal coated with gloss white paint. Luminaires with greater beam control use polished metal reflectors that are shaped to provide the desired beam and field angle. Reflectors can also shield the light source from direct view, thus reducing the possibility the luminaire will be a source of glare.

For recessed luminaires, the reflector can be finished at the ceiling in one of two ways – with or without a flange. The flange is a horizontal ring at the aperture of the reflector whose main purpose is to cover any irregularity in the edge of the adjacent ceiling tile or gypsum board. Most manufacturers offer the flange, or overlap flange, in two finishes. It can be the same color and finish as the reflector, often referred to as self-flanged, or it can be painted white. Some manufacturers also offer custom color paint at an additional cost.

The other flange finish is called flangeless. A flangeless reflector can be used with a gypsum board ceiling to minimize the aperture's appearance. It has a metal rim at the aperture, which is surrounded by a perforated metal ring or a ring of screen-like material. Once installed by the electricians, the gypsum board installers will use joint compound to cover the ring and seamlessly blend the gypsum board ceiling up to the aperture. Note that this type of flange requires more labor and is therefore more expensive to install than an overlap flange.

6.3.3 Beam control – lenses and diffusers

In addition to a reflector, some luminaires use a lens (refractor) or diffuser for additional beam control. Refractors and diffusers are described in more detail in Chapter 2.

6.3.4 Glare control – baffles and louvers

Luminaires that have a large aperture and/or that place the light source close to the aperture are more likely to be sources of glare. To minimize this possibility, some luminaires have one or more blades, baffles, or louvers at the aperture, as shown in Figure 6.2(c).

The words blade, louver, and baffle are not clearly defined and are used somewhat interchangeably. A blade generally describes one of a series of parallel vertical elements at the aperture of a luminaire that block direct view of the light source or diffuser. A baffle is similar to a blade, but the word usually describes two blades at a right angle to one another, which is also called a cross-baffle. Louver usually describes a set of blades in a dense grid (sometimes as tight as 1", or 25 mm, on center) set at right angles to one another. Track lights may use a louver where each cell has six sides, called a hex-cell louver. Some louvers have tall sides that are segments of a parabola to aid in controlling the light distribution, especially in linear fixtures, and are called parabolic louvers.

6.4 Luminaire Types

Once a designer begins to identify the types of luminaires needed and search for them, they will often search by luminaire type or function, and most manufacturer's websites provide a sorting of luminaires by function as one of the search options.

6.4.1 Downlight

A downlight (Figure 6.2) does exactly what its name suggests, directs light downward. The fixed, downward beam of light can be a cone (small aperture downlight), a pyramid (large aperture downlight such as a 2×4), or a wedge (linear downlight), and the luminaire can be suspended from the ceiling (pendant), attached to the ceiling (surface mount) or above the ceiling (recessed).

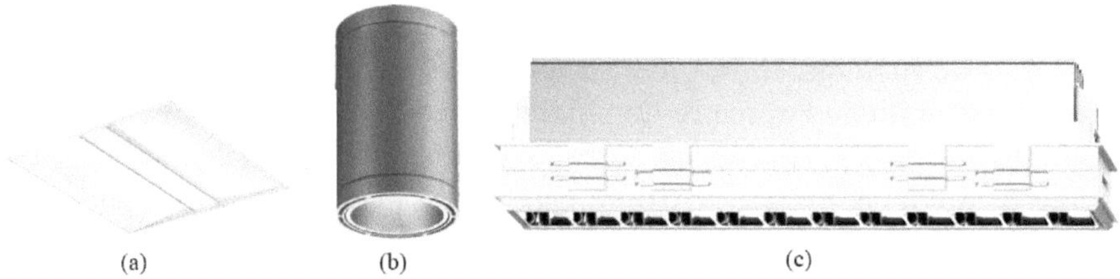

Figure 6.2 (a) Recessed 2 × 2 downlight. (b) Surface mount cylinder downlight. (c) Recessed multi-cell downlight (not to scale).
(a, b) Courtesy Acuity Brands Lighting, Inc.
(c) Courtesy Erco Lighting, Inc.

6.4.2 Accent or adjustable luminaire

An accent, or adjustable, luminaire emits a cone of light, but the direction and tilt of that cone are not fixed, (Figure 6.3). These luminaires are frequently

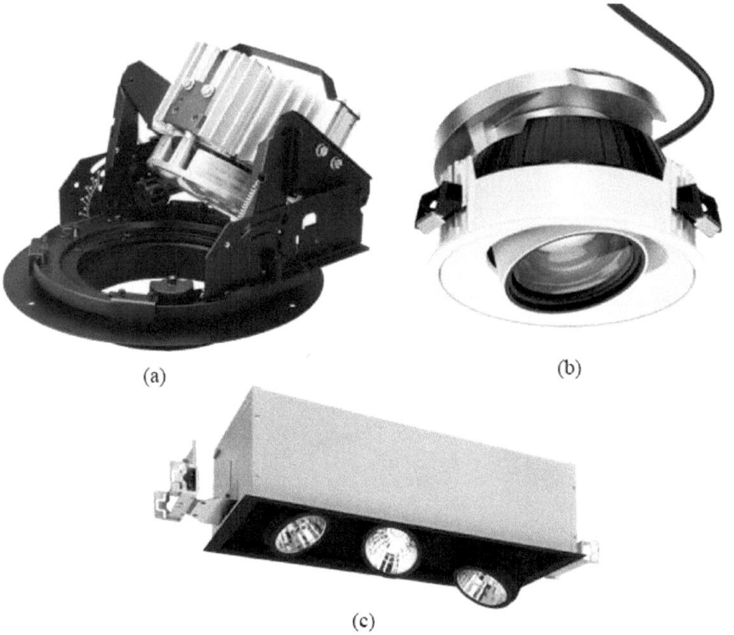

Figure 6.3 (a) Internal components of accent or adjustable luminaire. (b) Recessed adjustable luminaire from below. (c) Recessed adjustable multiple (not to scale).
(a, c) Courtesy of Acuity Brands Lighting, Inc.
(b) Courtesy of Erco Lighting, Inc.

used in retail, museum, and other display applications, because objects in the space need to be highlighted but their size, shape, and/or location changes over time. The ability to refocus luminaires provides the flexibility these applications require. Some luminaires have more than one adjustable module, permitting a single luminaire to illuminate more than one target. These luminaries are usually described as multiples.

In recessed accent luminaires, the rotation and tilt of the beam are usually manually adjustable, although motorized systems are available from a few manufacturers. Rotation of 355° to 360° and tilt of up to 30° from vertical are commonly available. Greater tilt tends to block a portion of the beam, so those fixtures are less common, and no manufacturer offers a tilt beyond 45°. In most cases the beam angle is field adjustable via replaceable lenses or reflectors, providing greater flexibility as the illuminated object changes size and location over time. The limited tilt range requires designers to examine the beam in a section drawing to confirm that the luminaire can illuminate its intended target.

6.4.3 Wall washer

A wall wash luminaire produces a fixed beam of light that, when used in groups and at the proper spacing, evenly illuminates a wall from floor to ceiling and edge to edge. Figure 6.4 is a section through a typical recessed lensed wall washer.

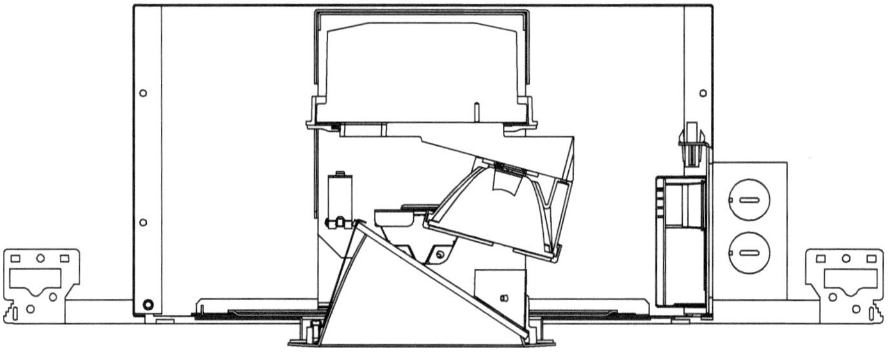

Figure 6.4 Section through a lensed wall wash luminaire. Courtesy of Acuity Brands Lighting, Inc.

6.4.3.1 Laying out wall washers

The optics of wall washers is similar across manufacturers, and their layout follows a few rules-of-thumb unless the manufacturer indicates otherwise.

1. For shorter walls, 12' (3.5 m) for example, the luminaires should be no closer to the wall than 2'-6" (0.75 m) and no farther from the wall than 4'-0" (1.25 m). For taller walls luminaires are typically set back from the wall at a distance of 0.3 to 0.35 of the wall height.
2. Once the distance from the wall is established it is also the on-center spacing of the luminaires, as shown in Figure 6.5.

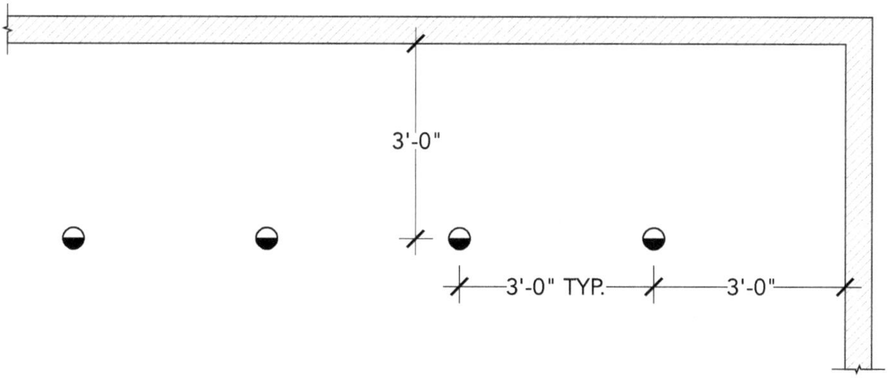

Figure 6.5 Wall washer spacing.

Making the on-center spacing of the luminaires greater than the offset from the wall is likely to result in visible scallops of light, which is undesirable in wall washing. Decreasing the spacing increases beam overlap, and therefore brightness, although it is more cost effective to increase brightness by selecting a higher lumen output than by using more luminaires.

6.4.4 Wall grazer

A wall washer is installed a short distance from the wall and delivers even lighting across the surface of the wall. A wall grazer is installed at the top or bottom of the wall (Figure 6.6) and skims light down or up the surface to create a gradient of light. Wall washers somewhat minimize wall texture. Wall grazers emphasize wall texture and are often used for that purpose.

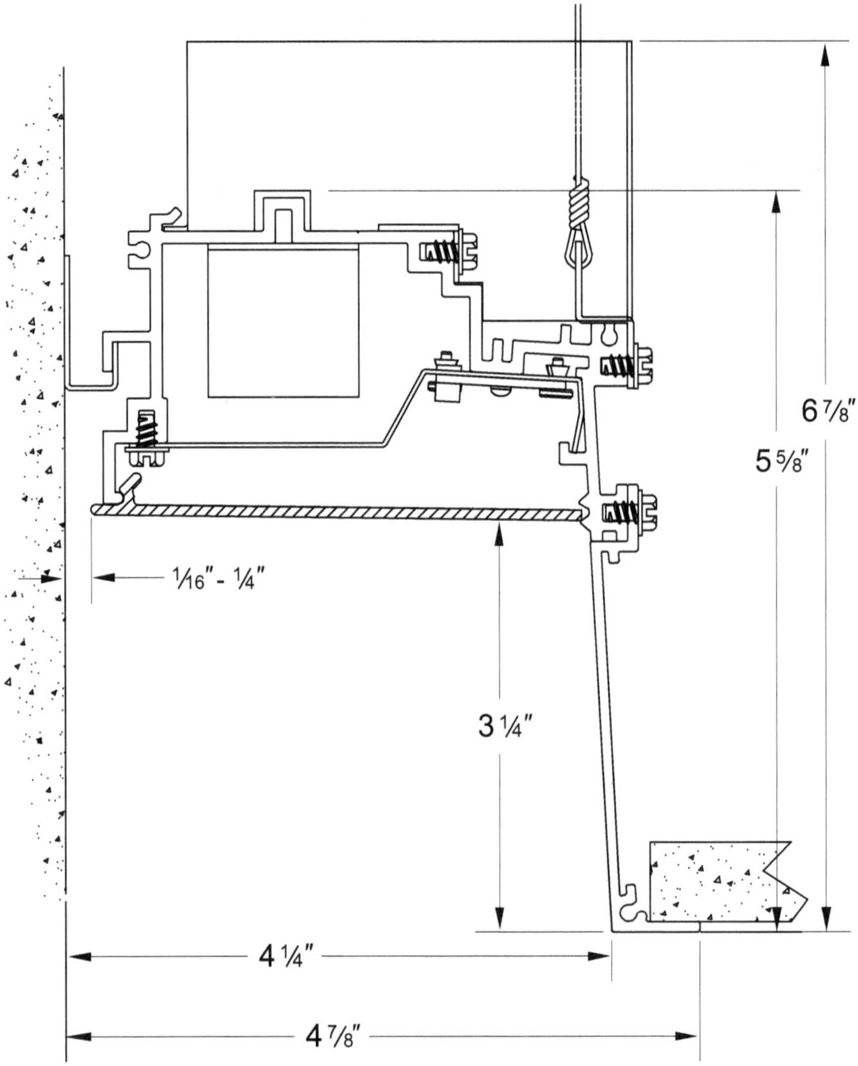

Figure 6.6 Section through a wall graze luminaire.
Courtesy of Acuity Brands Lighting, Inc.

6.4.5 Cove light

Cove lighting places a luminaire on an architectural shelf or recess below the ceiling, typically with a lip to prevent direct view of the luminaire. The light is directed toward the ceiling and can be used to create a decorative edging or to provide a soft, ambient light to increase room brightness. The purpose of the cove will dictate the luminaire and its output.

To light a room, or add meaningfully to room brightness, practitioners typically select a luminaire, such as the one shown in Figure 6.7 with high output and optics designed to throw light across the ceiling. Manufacturers of these types of luminaires typically include details for the size of the ledge and the optimal distance from the ceiling.

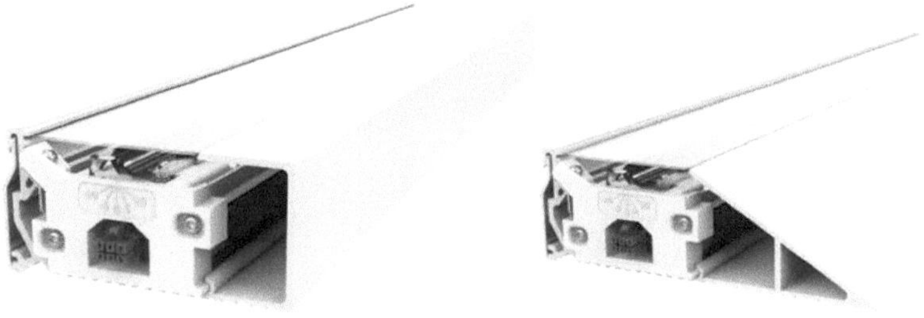

(a) (b)

Figure 6.7 Cove luminaires (a) in a rectangular extrusion and (b) in a knife-edge extrusion (not to scale).

Courtesy of Acuity Brands Lighting, Inc.

To create a decorative glow, practitioners often choose LED tape because it is relatively inexpensive, available in all CCTs and colors, and easy to install. The LED tape is often installed in an aluminum extrusion with a clear or frosted lens to protect it from damage and simplify cleaning.

6.4.6 Track and track heads

Consumers can purchase kits containing short lengths of track and a few track lights. Professional lighting designers specifying track systems must identify the track and each style of track light individually because they are not sold together or as kits.

6.4.6.1 Track

Track is generally an aluminum extrusion with integral copper bars to conduct electricity. Track can be recessed, surface, or pendant mounted and is typically available in white and black, although some manufacturers offer other colors. As with other recessed luminaires, recessed track can be flanged or flangeless.

While most people think of track as a conducting a single circuit of electricity so that all installed fixtures operate together, in fact track is available with up to four circuits to provide enhanced control (Figure 6.8).

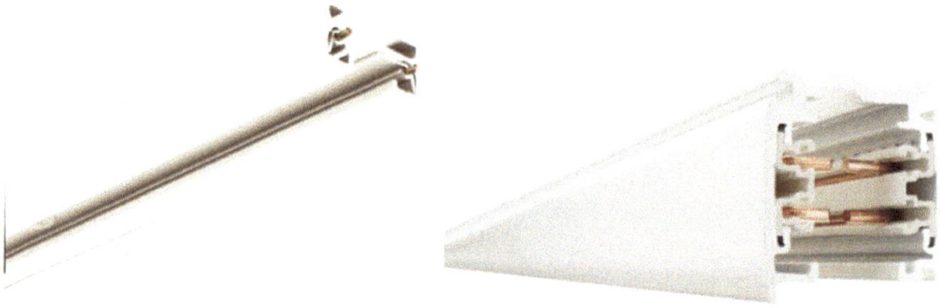

(a) (b)

Figure 6.8 (a) Single-circuit track. (b) Two-circuit track (not to scale). Courtesy of Acuity Brands Lighting, Inc.

6.4.6.2 Track heads

Track lights, or heads, are available in a wide range of purposes and styles to blend in with most interior designs. Optically they can be accent or spot lights (Figure 6.9) and wall washers. Some manufacturers offer decorative pendants, as well. Some track lights, especially those intended for museums

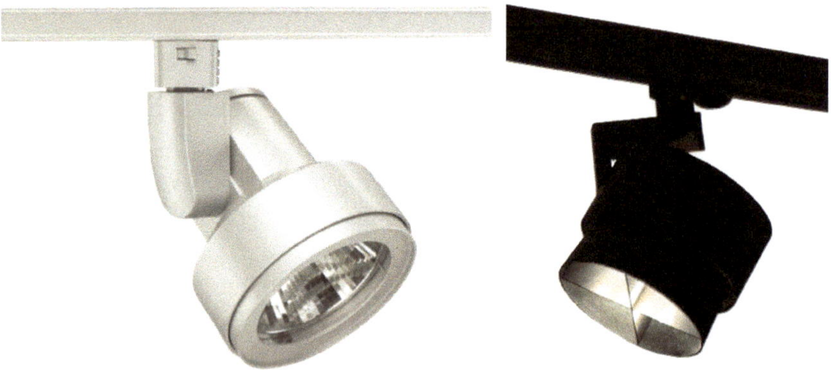

(a) (b)

Figure 6.9 (a) Track head. (b) Track head with cross baffle (not to scale).
(a) Courtesy of Acuity Brands Lighting, Inc.
(b) Courtesy of Erco Lighting, Inc.

and high-end retail, can hold up to three accessories that modify the beam, add color, or provide glare control.

6.4.6.3 LED tape

LEDs have made a new type of fixture possible – LED tape. As Figure 6.10 illustrates, the tape is a flexible strip with LEDs attached to its face at regular intervals. LED tape has become common in cove, under cabinet, under counter, and similar installations. LED tape is field cuttable, which can make installation easier, and some manufacturers offer side bending tape that can follow a curve. Many lighting practitioners specify an aluminum extrusion with a plastic lens (Figure 6.11) to protect the LEDs tape, hold it firmly in place, and improve diffusion of the light.

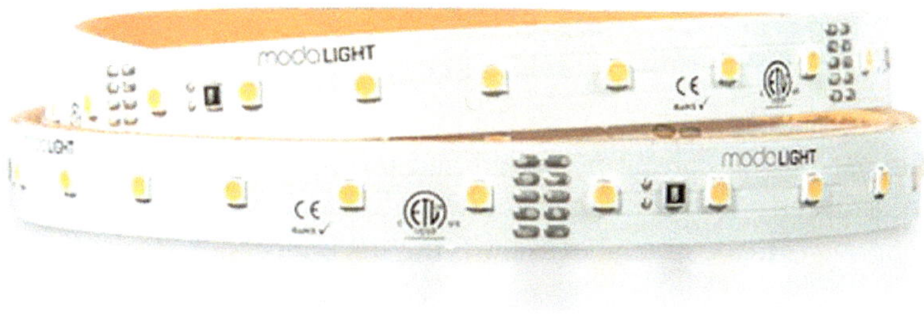

Figure 6.10 LED tape.
Courtesy of MODA Light

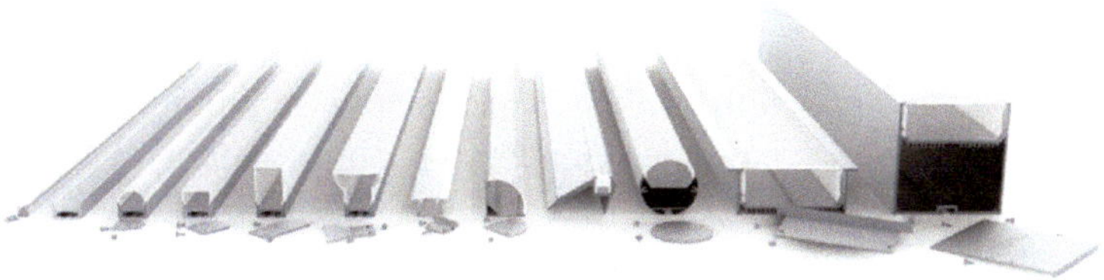

Figure 6.11 Extrusions for LED tape.
Courtesy of MODA Light

6.5 Retrofit Lamps and Kits

As discussed in Chapter 4, most lamp shapes originally developed for older lighting technologies are available today with LEDs as their light source. In incandescent luminaires, the retrofit process is as simple as removing the old lamp and inserting the new LED lamp, although choosing the correct shape, size, output, CCT, etc. is part of the retrofit process.

Ballasted luminaires, such as those using linear fluorescent lamps, can be upgraded with retrofit, or relight, LEDs in several ways. The fastest, easiest method is with an LED retrofit lamp, designated as Type A by UL, designed to operate on the luminaire's installed ballast. In some cases, these "lamps" are LED tape with an adhesive backing so they can be attached to the inside of the luminaire. Many of these retrofit lamps are not dimmable, even if the ballast is. Practitioners should confirm if dimming is required and, if so, be sure to select dimmable retrofit lamps.

The second method is to use line voltage retrofit lamps, which are designated as Type B by UL. These lamps require that the ballast be removed from the luminaire, or at least bypassed, by appropriately trained personnel before they are installed. A label must also be installed on the luminaire indicating a change to line voltage has been made.

Note that hybrid lamps are also available and are classified dual (Type A/B per UL). These lamps work with the existing compatible ballast. Later, when the ballast fails it can be bypassed and the lamp will work with line voltage wired directly to the fluorescent lamp holders (like Type B above). When this change is made, the label identifying the conversion to line voltage must be placed on the luminaire.

The third method uses a kit that includes lamps and a driver (Type C per UL). The driver replaces the luminaire's ballast, and the lamps are paired with the driver for optimal performance. A label must be placed on the luminaire indicating that a change to Type C has been made. In many cases, the drivers and lamps are dimmable. Practitioners should confirm that dimming is desired and match the driver to the existing dimming technology (forward phase, ELV, 0-10V, etc.). See Chapter 9 for information on dimming technologies.

6.6 Luminaire Installation Descriptions

Another method of organizing luminaires, both for designer search and manufacturer website purposes, is by how the luminaires are installed relative to their mounting surface.

6.6.1 Recessed

A recessed luminaire is one where the entire housing is above the ceiling, within a wall, or below the surface of the sidewalk, roadway, or ground.

The only visible components of the luminaire are the reflector and the lens or diffuser, if present. In most interior applications recessed luminaires are selected for their unobtrusiveness, leaving a clean ceiling, and not distracting occupants from the architectural and interior designs.

6.6.2 Semi-recessed

Most accent luminaires are fully recessed, but with some the adjustability comes from an "eyeball" that drops below the ceiling. Any luminaire that is mostly recessed but has components projecting below the ceiling is described as semi-recessed.

6.6.3 Surface mount

Surface mount luminaires are those that are affixed to the surface of the ceiling or wall. Surface mount luminaires include downlights, track and track heads, and wall sconces.

6.6.4 Pendant

Pendant describes any luminaire that is hanging from the ceiling. Pendant luminaires may be decorative, as with chandeliers, or functional as with linear pendants illuminating an office.

6.6.5 Poles and bollards

Tall exterior luminaires such as those used to light roadways, parking lots, parks, etc. are described as pole mount. Short luminaires, such as those used to light walkways, are called bollards.

6.7 Luminaire Classifications

6.7.1 CIE distribution categories

The CIE luminaire classification system goes back to at least the early 1940s. Today, the categories are defined in the CIE International Lighting Vocabulary (ILV). The six categories for indoor luminaires are based on the proportion of light they emit upward and downward. These distribution categories are well understood and are used throughout the lighting industry (DiLaura et al., 2011). See Figure 6.12.

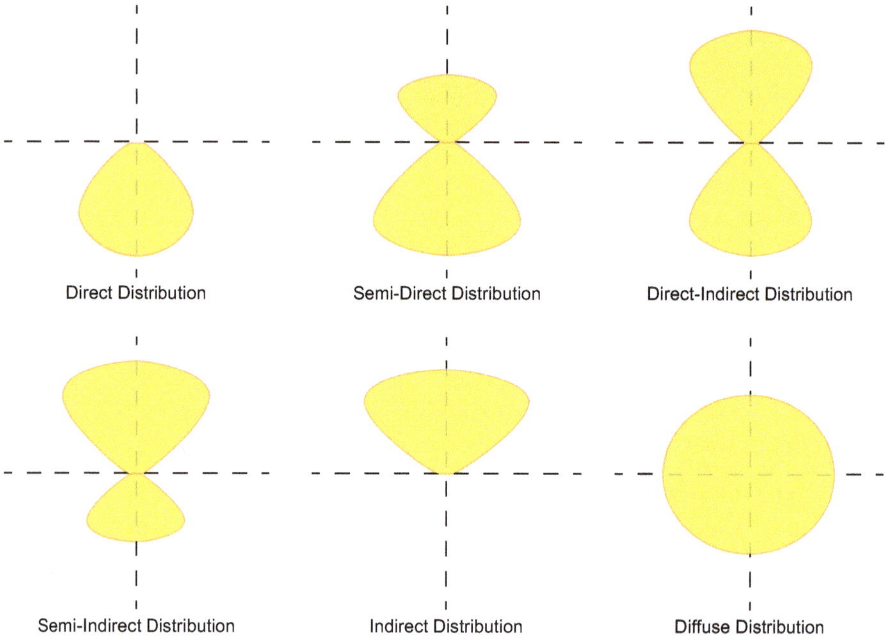

Figure 6.12 CIE distribution categories.

6.7.1.1 Direct

Direct distribution describes a luminaire that emits between 100% and 90% of its light directly toward the plane of calculation. Downlights have a direct distribution, as do wall washers, wall grazers, and track heads.

6.7.1.2 Semi-direct

Semi-direct describes luminaires with a distribution where at least 60% of the light is directed downward and up to 40% of the light is directed upward. Since they have an upward component to their distribution, they are pendant-mounted or, in rare cases, wall-mounted.

6.7.1.3 Direct-indirect

Direct-indirect describes pendant luminaires where both the upward and downward distribution are between 40% and 60% of the output.

6.7.1.4 Semi-indirect

Semi-indirect luminaires send at least 60% of their light upward and up to 40% of the light downward.

6.7.1.5 Indirect

Indirect luminaires emit between 100% and 90% of their light upward toward the ceiling, thus indirectly lighting the space by bouncing the light off the ceiling.

6.7.1.6 Diffuse

Diffuse luminaires emit light more or less equally in all directions. A chandelier is an example of a diffuse luminaire.

6.7.2 Outdoor luminaire classifications

6.7.2.1 IES cutoff classification

One of the earliest outdoor classification systems was developed by the IES in 1963 and was based on the amount of light emitted above 80°. The goal was to limit light that was wasted and that contributed to sky glow. The categories are described in Table 6.1:

Table 6.1 IES cutoff classifications.

IES cutoff classification	Percentage of lumens distributed above 80°
Full cutoff	0%
Cutoff	0% to 20%
Semi-cutoff	20% to 40%
Non-cutoff	Over 40%

This system is simple and easy to use; however, a newer IES luminaire classification system using BUG ratings has replaced cutoff classification for most manufacturers and practitioners.

6.7.2.2 IES luminaire classification system and BUG ratings

The IES luminaire classification system analyses the percentage of lumens that are emitted as forward light (directed to the ground in front of the luminaire), back light (light emitted behind the luminaire), and up light (light directed above the luminaire). Each of the three areas is subdivided into several smaller angular ranges as shown in Figure 6.13, and lumens emitted within each range measured. The measurements are used to assess backlight, up light, and glare (a combination of back light and front light measurements) categorized from 0 (the most restrictive) to 5 (the most lenient) in each category. The three category values (B0 to B5, U0 to U5, and G0 to G5) are combined to generate a BUG (Backlight, Uplight, light directed toward the Ground) rating for each luminaire measured (IES Testing Procedures Committee, 2020).

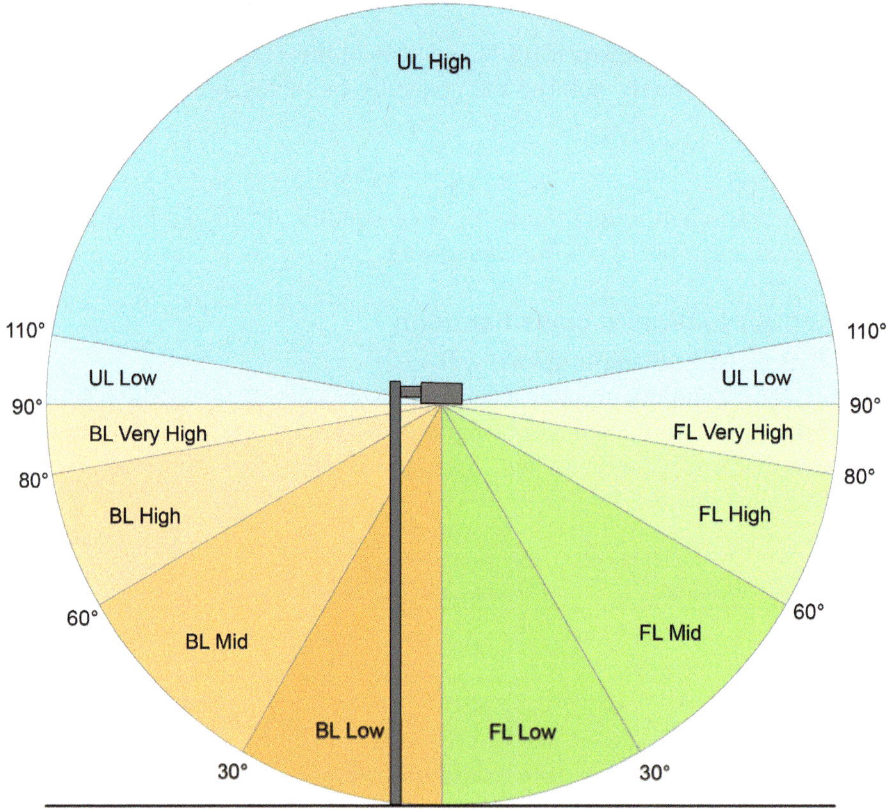

Figure 6.13 BUG zones.

6.7.2.3 IES distribution classifications

The IES distribution classification system describes the shape of the illuminated area on the ground and is used for roadway and area lighting (DiLaura et al., 2011). This system has six categories (Figure 6.14) that are described as follows:

- Type I: Narrow, symmetric distribution with highest intensity at nadir (straight down)
- Type II: Wider distribution than Type I with highest intensity between 10° and 20°
- Type III: Wide distribution with highest intensity between 25° and 35°
- Type IV: Widest distribution with highest intensity above 35°

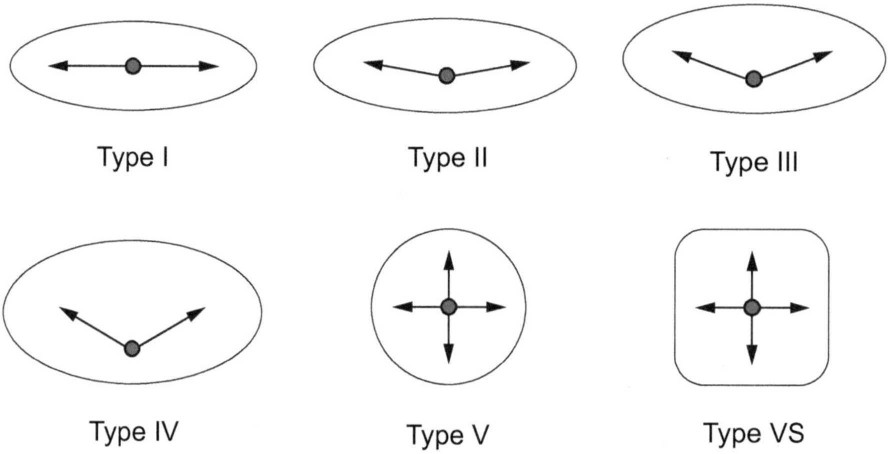

Figure 6.14 IES outdoor illumination distribution patterns.

- Type V: Symmetrical beam producing a circular pattern on the ground
- Type VS: Symmetrical beam producing nearly square pattern on the ground

6.7.2.4 NEMA beam angle classifications

NEMA's beam angle system has seven categories based on the luminaire's field angle (the angle across the beam at which intensity has fallen to 10% of center beam brightness), shown in Table 6.2, and is used for sports lighting and floodlighting. A single number indicates a circular beam, while two numbers indicate an elliptical beam. For example, a 25° × 60° beam would be described as NEMA 2×4.

Table 6.2 NEMA beam spread classifications.

Field angle	NEMA type	Description
10° to 18°	1	Very narrow
18° to 29°	2	Narrow
29° to 46°	3	Medium narrow
46° to 70°	4	Medium
70° to 100°	5	Medium wide
100° to 130°	6	Wide
130° and up	7	Very wide

References

DiLaura, D.L., Houser, K.W., Mistrick, R.G., Steffy, G.R. (2011). *The Lighting Handbook, 10th Ed.* New York: Illuminating Engineering Society.

IES Testing Procedures Committee (2020). *ANSI/IES TM-15-20 Technical Memorandum: Luminaire Classification System for Outdoor Luminaires.* New York: Illuminating Engineering Society.

Livingston, Jason (2022). *Designing with Light: The Art, Science, and Practice of Architectural Lighting, 2nd Ed.* Hoboken: John Wiley & Sons, Inc.

7

Daylighting

Abstract

Daylight is plentiful, high-quality, and free. It can also be overwhelmingly bright and cause unacceptable glare. This chapter introduces the benefits of daylighting as well as the challenges. It explains the role of the architect, interior designer, and lighting practitioner in successfully implementing daylighting. It discusses the two main approaches to daylighting – side lighting and top lighting. Examples are given of energy conservation code requirements for daylighting, and the calculation of daylight zones adjacent to windows, under clerestories, and under skylights. The chapter ends with a review of light loss factors that must be included in any illuminance calculations to ensure that the calculations are accurate.

Keywords: Daylighting, daylight harvesting, skylight, clerestory, solar heat gain, glazing, fenestration, light shelf, roof monitor, visible transmission, side lighting, top lighting.

7.1 Introduction

Daylighting refers to the practice of using light from the sun and sky to illuminate building interiors. For centuries, daylighting was integral to an architect's work because other sources of light (candles, torches, etc.) were dim, dirty, and expensive. With the invention of electric lighting, daylighting practices seemed less necessary, and architects designed ever larger floorplates that placed most occupants far from windows. In the late twentieth and early twenty-first century, ever tightening power allowances in the energy conservation codes renewed interest in daylighting. Energy conservation, combined with the realization that daylighting provides many benefits in addition to free illumination, has added daylighting to the lighting practitioner's areas of expertise.

In daylit spaces, electric lighting plays a supporting and supplementary role that fills the gap (if any) between daylight illuminance and the lighting design's target illuminance. While it is not possible, or desirable, to incorporate daylighting into every building, daylighting should be considered where appropriate because of the potential for considerable energy savings and occupant satisfaction.

A nomenclature note: Glass assemblies in building facades are often referred to as glazing or fenestration. Windows and clerestories are sometimes described as vertical glazing or fenestration. Skylights are sometimes described as horizontal glazing or fenestration. This chapter will use the term glazing.

7.2 Benefits of using Daylighting

Incorporating daylighting into architectural and lighting designs provides many benefits to owners and occupants. It provides light that is free, has high color fidelity (provided it is not overly modified by glazing), is interesting in its variations over time, and provides a connection to the outdoors and the passage of time. These can combine to improve the experience of owners and occupants in many ways, including:

7.2.1 Reduced energy costs

Daylight doesn't eliminate the need for electric lighting. Except in very shallow floorplates, it can't illuminate the entire floor and, of course, it isn't available at night. However, using photosensors and lighting controls to reduce the brightness of electric light when daylight is present (referred to as daylight harvesting) can result in energy savings of 40% or more in side-lit daylight zones and 50% or more in top-lit zones (IES Daylighting Committee, 2020).

7.2.2 Increases in occupant satisfaction

Windows and skylights introduce variations in light that come with changes in weather, time of day, and seasons. These changes provide a connection to the outside and an indication of the passing of time, which most people appreciate. Occupants express greater satisfaction in spaces with daylighting and views to the outside. In addition, the variations in patterns of light and shadow add dynamic visual interest to a space (Boyce, et al., 2003; Jamrozik, et al., 2019; IES Daylighting Committee, 2020).

7.2.3 Increased retail sales

As noted in Chapter 1, multiple studies have found that retail stores with daylight provided by skylights have retail sales 10% (Ellis, 2007) to 40% (Heschong, 2003) higher than similar stores without daylight.

7.2.4 Enhanced employee well-being and productivity

Studies have shown that workers in daylit offices report better sleep quality and quantity, and a more positive mood in the workplace (Figueiro & Rea, 2016). Worker's cognitive function improves in spaces with access to daylight and views, and workers rate window access as one of the most important improvements that could be made in windowless spaces (Jamrozik, et al., 2019).

7.2.5 Increased student performance

Daylight in classrooms has been shown to positively impact student retention of information and test scores (Heschong, et al., 2013; Baloch, et al., 2021).

7.3 Challenges of using Daylighting

While there are many benefits to adding daylighting to a space or building, it requires care to avoid simultaneous draw backs, which include the following (IES Daylighting Committee, 2020).

7.3.1 Solar heat gain

Daylight from the sky is a highly desirable light source. Direct sunlight, on the other hand, is not, for several reasons. The first is the high intensity of light, 10,000 fc or 100,000 lx, gets converted to heat when absorbed by interior surfaces. In addition, 54% of the energy in direct sunlight is outside of the visible spectrum in the infrared and ultraviolet regions. Skylights that don't block this energy allow even more heat into the building. Heat added to the building by daylight and direct sun is referred to as solar heat gain. If the daylighting system is not carefully designed, the costs of a larger HVAC (heating, ventilation, air conditioning) system to handle solar heat gain and the energy used to remove it from the building can easily exceed the energy savings of reduced electric lighting.

7.3.2 Building heat loss

Skylights and windows are usually less efficient insulators than the walls and roofs that surround them. As a result, building and/or energy conservation codes set a maximum area for skylights. For example, *ANSI/ASHRAE/IES Standard 90.1* limits skylights, with few exceptions, to 3% of the total roof area. It also sets a maximum U-factor (a measure of heat loss through a widow or skylight) and maximum solar heat gain coefficient (a measure of how well a window or skylight blocks heat from direct sunlight) for skylight assemblies.

7.3.3 Glare

Another potential problem with direct sunlight is glare, which is also discussed in Chapter 10. Direct sunlight can be a source of direct glare for workers that face windows and have a direct view of the sun during parts of the day. The brightness of the sun is annoying at best, and at worst can cause eye strain, headaches, and make visual tasks difficult or impossible because of the extreme contrast between the brightness of the sun and the illuminance on tasks.

Even if students or workers don't have a direct view of the sun, sunlight landing on walls, white boards, displays, and desktop tasks can create reflected glare that make learning or work difficult.

7.3.4 Excessive brightness

It would be easy to assume that if one skylight provides some light, more skylights will provide more light and save more energy. That is true up to a point, but illumination recommendations for interior visual tasks are nearly all below 100 fc (1000 lx), which is 20% of the brightness of light from the sky and 1% of the brightness of direct sunlight. Care must be taken to size and space skylights, clerestories, and windows to admit a controlled amount of daylight to avoid excess illumination and the possibility of glare.

7.4 The Role of Architects and Lighting Practitioners in Daylighting

Daylighting is not a lighting strategy that is the exclusive domain of the lighting practitioner. The architect and interior designer are key collaborators in a successfully designed and implemented daylighting scheme. The owner must also agree, because successful daylighting has implications for the building orientation and façade, and the interior design.

The architect has primary responsibility for the orientation of the building on the site. Aligning the long axis of the building with east/west maximizes building faces that receive consistent sunlight (the southern face in the northern hemisphere) and consistent daylight (the northern face in the northern hemisphere). Similarly, the long axis of rooms should be parallel to the façade. An analysis of maximum and minimum solar angles will reveal the depth that sunlight will penetrate the building, and the potential for direct glare. Working with other team members (especially the lighting practitioner) and the building owner, the design team can develop strategies to maximize daylight and minimize solar heat gain and glare.

The interior designer should select materials and finishes that have high reflectance values, which helps daylight penetrate deeper into the building. Workstation partitions should be low so they don't block the daylight. Likewise, the layout of private and open offices and other types of spaces should be arranged and designed to minimize barriers to daylight. For example, often private offices, with their important occupants, are placed at the perimeter of the building to provide views. However, since private offices often have full-height, opaque walls, they block daylight from the interior of the building. Better strategies include moving private offices to the interior of the building and/or using transparent or translucent materials for partition walls so light can pass through them.

The lighting practitioner, who is likely to be the team member most aware of the advantages and pitfalls of daylighting, can advise the architect and interior designer throughout the design process. Collaboration on issues from methods of directing light deeper into the building, to shading schemes that block direct sunlight, to interior design choices help to keep the design team focused on supporting the daylighting strategy to maximize its effectiveness and reduce energy consumption.

7.5 Daylighting Techniques

In general, there are two approaches to daylighting: side lighting and top lighting. Side lighting describes light that enters through windows or clerestories. Top lighting describes light that enters through skylights or rooftop monitors. Top lighting is a more effective solution, since it can cover a larger floor area more evenly. Side lighting, however, is the more common approach. After all, only the top floor of any building has a roof that can use skylights, but every floor has exterior walls with windows.

7.5.1 Side lighting

Side lighting is using daylight that enters a space through windows or clerestories as a source of illumination. With side lighting, the rule of thumb is that daylight may penetrate the room up to 2× the distance from the floor to the top of the glazing, although it doesn't mean a useful amount of light will always reach that far into a space. Larger windows will admit more light than smaller windows. High or tall windows light deeper into a space than low or short windows. The applicable energy conservation code will specify the calculation to be used in establishing the area that is daylit and the associated lighting control zones.

Figure 7.1 shows that when using a window alone there can be a rapid drop-off in the daylight illuminance level. One method of getting daylight deeper into the room is with a light shelf (Figure 7.2). A light shelf is a horizontal surface that reflects incoming daylight and sunlight onto the ceiling, where it is diffused and reflected deeper into the room. A light shelf divides the window into two halves. The upper half is glazed to admit light, while the lower half is glazed for views.

The top of the light shelf is a highly reflective surface, although specular materials should be avoided because they can create bright ceiling areas that may be a source of glare. A light shelf can be placed inside, through, or outside of the window. Light shelves that are outside the window also shade the lower glazing, reducing heat gain from direct sunlight. Light shelves are

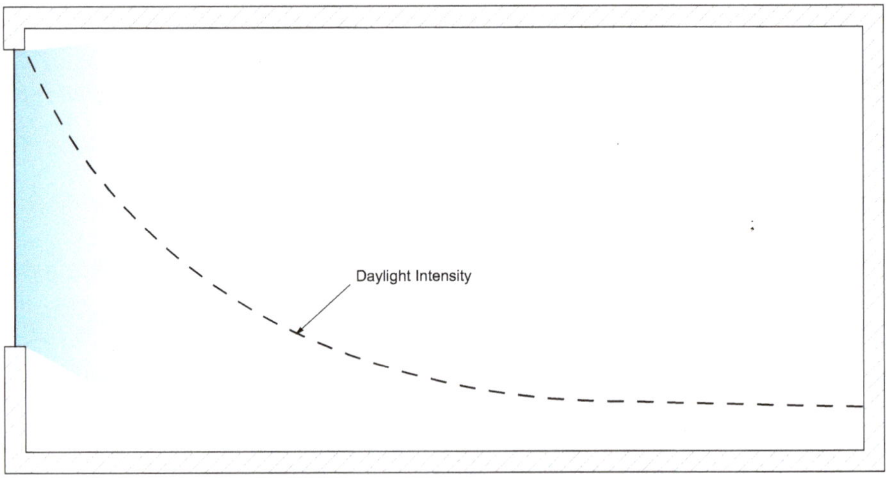

Figure 7.1 Daylight distribution through a window.

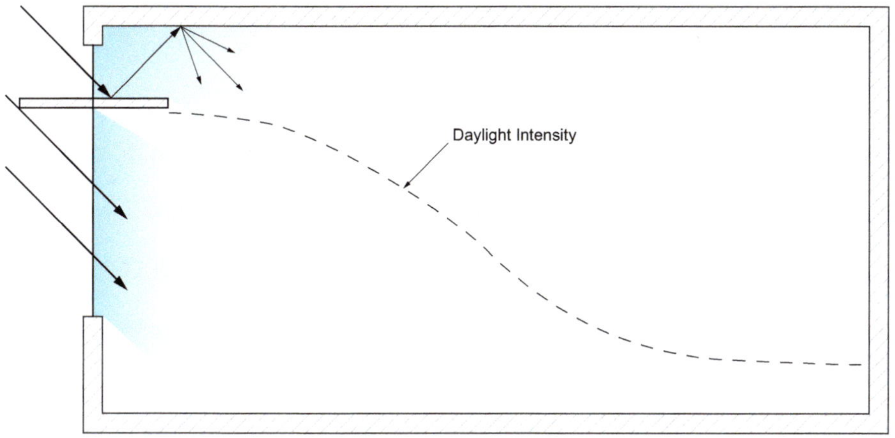

Figure 7.2 Daylight distribution using a light shelf.

typically installed at 7'-6" (2.3 m) or higher, so ceiling heights of 9' (2.75 m) or higher are required.

7.5.2 Top lighting

Top lighting is using skylights or roof monitors to introduce daylight into an interior space. Skylights can be flat, gabled, domed, or another shape. Because skylights can admit direct sunlight into a space, they often use a frosted or white material to diffuse the sunlight and widen the distribution of light. Dome and barrel vault shaped skylights collect more daylight than do flat skylights. Sloping the sides of the skylight well also provides broader distribution and softens the contrast between the ceiling and the skylight. Adding a diffuser to the bottom of the skylight creates the broadest distribution (Figure 7.3). As noted earlier, skylight assemblies must be evaluated not just for their ability to allow light into a building, but for their insulating properties as well.

When using skylights to illuminate a large portion of a building, or an entire floor, the skylights should be evenly spaced and should cover between 2% and 6% of the roof to provide even illumination. A higher quantity of smaller skylights produces more even illumination than a smaller number of large skylights, even if the total area of skylights is the same. To provide acceptable uniformity, the spacing between skylights should not exceed 1.5 times the height of the space.

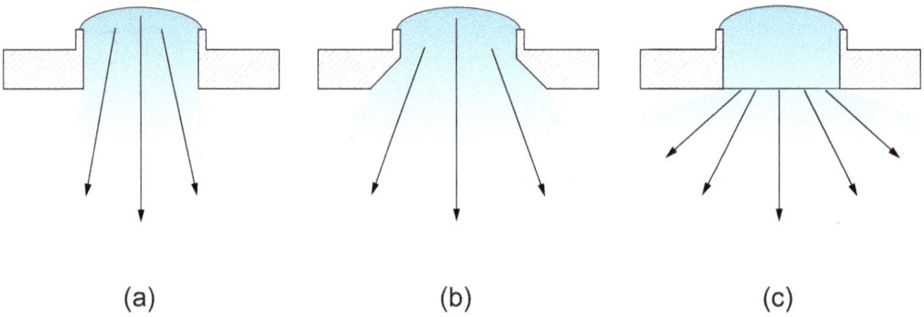

Figure 7.3 Skylight distribution. (a) Straight-sided skylight. (b) Slope-sided skylight. (c) Skylight with a bottom fiffuser.

Although less common, the light admitted by skylights can be controlled by the addition of manual or motorized vanes or louvers. These will block some of the light as they are rotated from vertical to horizontal, and can be used to maintain consistent illuminance over the course of the day.

Sawtooth roof monitors and clerestories fall between skylights, which admit light from above, and windows, which admit light from the side. The light from sawtooth monitors and clerestories is admitted at, or near, the ceiling but has a partial sideways distribution. Although they can be oriented in any direction, they are often designed to face north in the northern hemisphere and south in the southern hemisphere. This admits the most consistent and evenly distributed daylight and minimizes or eliminates direct sunlight.

Another top lighting approach uses "pop-up" roof monitors. These monitors have glazing on more than one side. The additional glazing admits more daylight than clerestories, and the resulting light is less directional. However, they can have the added complication of admitting direct sunlight if not carefully shielded or shaded.

7.6 Daylighting in Energy Conservation Codes

Daylight is a free source of light that can greatly reduce the need for electric light during the day. To ensure buildings take advantage of daylight, energy conservation codes set requirements for lighting controls in areas that receive daylight. For example, the *International Energy Conservation Code* has the following requirements:

1. Luminaires in daylight zones must be controlled via photosensors and must be controlled separately from luminaires outside of daylight zones.

2. Daylight zones, with some exceptions and modifications, are calculated as:

    a. Daylight zones adjacent to windows and clerestories are determined as follows: The width is calculated as the width of the window plus 2 ft (0.6 m) on each side of the window. The depth is equal to the distance from the floor to the top of the window (Figure 7.4).
    b. Daylight zones under rooftop monitors are determined as follows: The width is calculated as the width of the rooftop monitor plus one quarter of the height from the floor to the base of the glazing. The depth is equal to the distance from the floor to the bottom of the glazing (Figure 7.5).

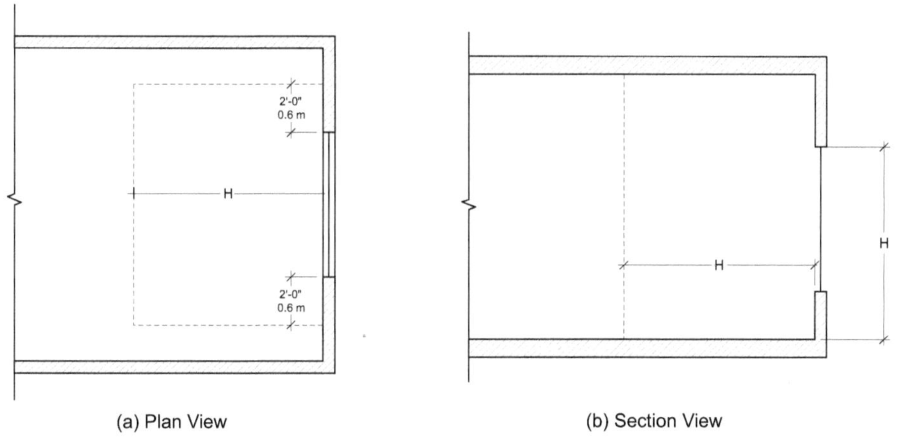

Figure 7.4 Daylighting side-lit zone adjacent to a window. (a) Plan view. (b) Section view.

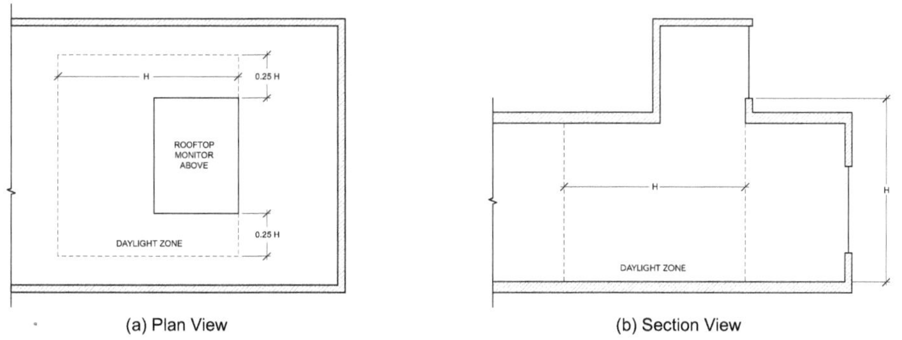

Figure 7.5 Daylighting zone under a rooftop monitor. (a) Plan view. (b) Section view.

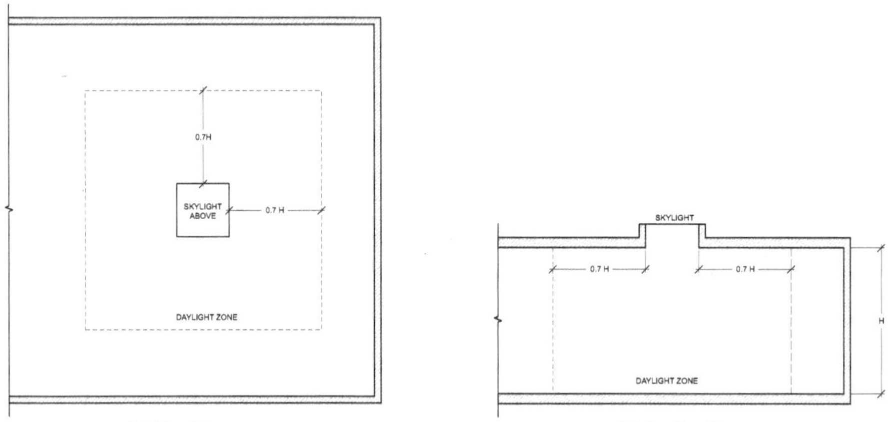

Figure 7.6 Daylighting zone under a skylight. (a) Plan view. (b) Section view.

c. Daylight zones under skylights are determined as follows: The width and depth are both the size of the skylight plus 70% of the ceiling height of the space (Figure 7.6).

7.7 Daylighting Design and Calculation Software Programs

Many of the illuminance calculation software packages used by lighting practitioners to calculate interior light levels (see Chapter 8) also are used for daylighting. In each case, the building's longitude and latitude is entered into the software along with the north/south orientation and the date and time to be studied.

Typically, at least three calculations are performed: the winter solstice, when the sun is at its lowest angle (in the northern hemisphere); the summer solstice when the sun is at its highest angle; and the vernal or autumnal equinox. This permits the design team to examine the results of the architectural, interior, daylighting, and electric lighting design choices at the two extremes and an intermediate condition.

The calculation software, depending on its capabilities, can then display illuminance data and color or black and white renderings. As the industry transitions to building information modelling (BIM) software the detail presented in renderings is drawn directly from the BIM model rather than a separately created model in the stand-alone calculation software. This often results in more detailed and accurate calculations, as well as more realistic renderings for clients.

7.7.1 Light loss factors in daylighting calculations

As with interior illuminance calculations, there are factors to be considered with daylighting calculations that reduce the average maintained illuminance.

- Glazing transmittance. There are few window and skylight models available for illuminance calculation software. It can be quick and easy to simply make a hole in the building model that is the same size and shape as the skylight. However, most glazing assemblies have treatments that block some portion of the visible spectrum. Light that passes through the glazing is the visible transmission, is measured as a percentage, and is abbreviated VT%. Skylights with white glazing may have a VT% of 40 or less, while clear low emissivity glass (which is treated to limit heat loss, often by reflecting IR back into the room) may have a VT% of 80. Accurate calculations require the inclusion of this light loss factor.
- Mullions. Window systems include framing and/or mullions that are, of course, opaque. It is not necessary to model every one of them. Lighting practitioners can calculate the percentage of the opening that is occluded by mullions and add that to the light loss factor calculation.
- Dirt depreciation. Atmospheric dirt builds up on windows and skylights. While there aren't good studies providing clear guidance, the IES currently recommends a dirt depreciation light loss factor of between 0.8 and 0.9 for skylights and 0.9 to 0.95 for windows (IES Daylighting Committee, 2020).
- Exterior factors. Most programs default to a flat exterior that is open to the sky, so the landscaping or other features may need to be added to the model for accurate calculations. For example, if a building is fronted by a line of shade trees or if one face is partially shaded by an adjacent building, the model and calculations should include the resulting reduction in daylight.

References

ASHRAE Standing Standard Project Committee 90.1 (2016). *ANSI/ASHRAE/IES Standard 90.1-2022 Energy Standard for Buildings Except Low-Riser Residential Buildings*. Atlanta, GA: ASHRAE.

Baloch R.M., Maesano C.N., Christoffersen J., Mandin C., Csobod E., Fernandes E.O., Annesi-Maesano I. (2021) Daylight and School Performance in European Schoolchildren. *Int J Environ Res Public Health.*;18(1):258. doi:10.3390/ijerph18010258.

Boyce, P., Hunter, C., Howlett, O. (2003) *The Benefits of Daylight Through Windows*. Troy, NY: Lighting Research Center, Rensselaer Polytechnic Institute.

Daylighting (2016). Availavle at: https://www.wbdg.org/resources/daylighting. [accessed April 11, 2024].

Ellis, George (2007). *Let the Sun Shine In*. Supermarket News. Available at: https://www.supermarketnews.com/retail-amp-financial/let-sun-shine [accessed January 22, 2024].

Figueiro M., Rea M. (2016). Office lighting and personal light exposures in two seasons: Impact on sleep and mood. *Lighting Research & Technology*. 2016;48(3):352-364. doi:10.1177/1477153514564098

Heschong, Lisa (2003). *Daylight and Retail Sales*. California Energy Commission. Available at: https://newbuildings.org/wp-content/uploads/2015/11/A-5_Daylgt_Retail_2.3.71.pdf [accessed January 22, 2004].

Heschong, L., Wright, R., Okura, S. (2013). Daylighting Impacts on Human Performance in School. *Journal of the Illuminating Engineering Society*. 31. 101-114. 10.1080/00994480.2002.10748396.

IES Daylighting Committee (2020). *ANSI/IES LP-3-20 Lighting Practice: Designing and Specifying Daylighting for Buildings*. New York: Illuminating Engineering Society.

International Energy Conservation Code Committee (2018). *International Energy Conservation Code*. Washington DC: International Code Council.

Jamrozik, A., Clements, N., Hasan, S.S., Zhao, J., Zhang, R., Campanella, C., Loftness, V., Porter, P., Ly, S., Wang, S., Bauer, B. (2019). Access to Daylight and View in an Office Improves Cognitive Performance and Satisfaction and Reduces Eyestrain: A Controlled Crossover Study. *Building and Environment*. Vol. 165.

8

Light Quantity and Calculations

Abstract

Brightness is a key component of any lighting design. This chapter begins by describing the units used to measure light intensity both at the source and landing on surfaces. Two calculations are explained with examples, one used for brightness at a point and one for average brightness across a room. Computer calculation software is also reviewed. The chapter closes by illustrating and explaining photometric graphics and photometric data files provided by luminaire manufacturers.

Keywords: Lumen, lux, footcandle, luminance, illuminance, coefficient of utilization, maintained illuminance, room cavity ratio, reflectance, light loss factors, lamp lumen depreciation, luminaire dirt depreciation, room surface dirt depreciation, voltage factor, thermal factor, ballast factor, reflector factor, spacing criterion, reflectance, point method, lumen method.

8.1 Introduction

During the design process for nearly every project there comes a time when the question of brightness arises. "How bright are the luminaires?" "How bright are the desktops, the walls, and the displays?" "How much brightness is needed?" In some cases, the answer is flexible and depends on the project, the owner, and the designer's goals. In others, there may be minimum or maximum light levels that must be adhered to. In any case, the calculations described in this chapter allow the lighting practitioner to assess the design for compliance with the target light levels, and/or calculate the light level that will be delivered by the design.

8.2 Units of Measurement

When measuring light, the units of measurement are specific to what is being measured: luminance (brightness of light at the source); illuminance (brightness of light falling on a surface) or reflectance (brightness of light reflected off a surface). Some of the units are the same in the U.S. Customary and International System of Units (SI) systems, while others are not.

8.2.1 Luminance

Luminance is the intensity of light emitted by a light source or reflected from an illuminated surface. Light sources and glowing surfaces are described as being luminous.

The lumen (lm) is the unit of measurement of the directional light emitted within one steradian. (A steradian is a 3-D angle, with its vertex at the center of a sphere, that encompasses a circle on the sphere whose area is the square of the radius. There are 4π or 12.57 steradians in a sphere.) In practice, when a manufacturer provides the lumen output of a lamp or luminaire, they are providing the total number of lumens generated, not just the lumens emitted in a single steradian. Note that some luminaire manufactures will list the "lumens" as the total output of the LED and separately list the "delivered lumens" as the output of the luminaire because light is lost as it is absorbed by the luminaire's components, such as the reflector or lens.

The candela (cd) is the intensity of a light source in a direction and is equal to one lumen per steradian. This measure is typically used to describe the intensity of light at the center of a beam and may be the only luminance measure given for directional lamps. A light source with a luminance of 1 lumen emits 12.57 candelas (cd). Although the candela officially replaced the candlepower (cp) in 1948, some manufacturers still list candlepower instead of candelas in their documentation. The difference is minor (1 cp = 0.98 cd), so to simplify calculations lighting specifiers treat them as being equal.

Light emitted from or reflected off a surface is treated as light from a plane not a point. The U.S. Customary unit of measurement is the candela per square foot (cd/ft^2), while the SI unit is the candela per square meter (cd/m^2).

8.2.2 Illuminance

When light lands on a surface that surface is illuminated. This is also described as illuminance or light incident upon a surface. Calculating the illuminance and understanding the result is an important aspect of lighting design. There is over a century of research into the illuminance required for the effective performance of a wide range of visual tasks. One of the lighting designer's responsibilities is to understand the visual tasks that are to be performed in a space and use calculations to confirm that the design is providing the appropriate illuminance. There are two units of measurement for illuminance (Figure 8.1).

A footcandle (fc) is the light illuminating a surface of one square foot that is one foot away from a light source of 1 lumen and is the U.S. Customary unit of illuminance.

The lux (lx) is the light illuminating a surface of one square meter that is one meter from a light source of 1 lumen and is the SI unit of illuminance.

In both cases part of the definition is that the light is evenly distributed (Rea, 2000). This will become an important consideration in some calculations discussed later in this chapter.

Conversion between the two units is as follows:

$$1\text{fc} = 10.76 \text{ lx and } 1 \text{ lx} = 0.09 \text{ fc}.$$

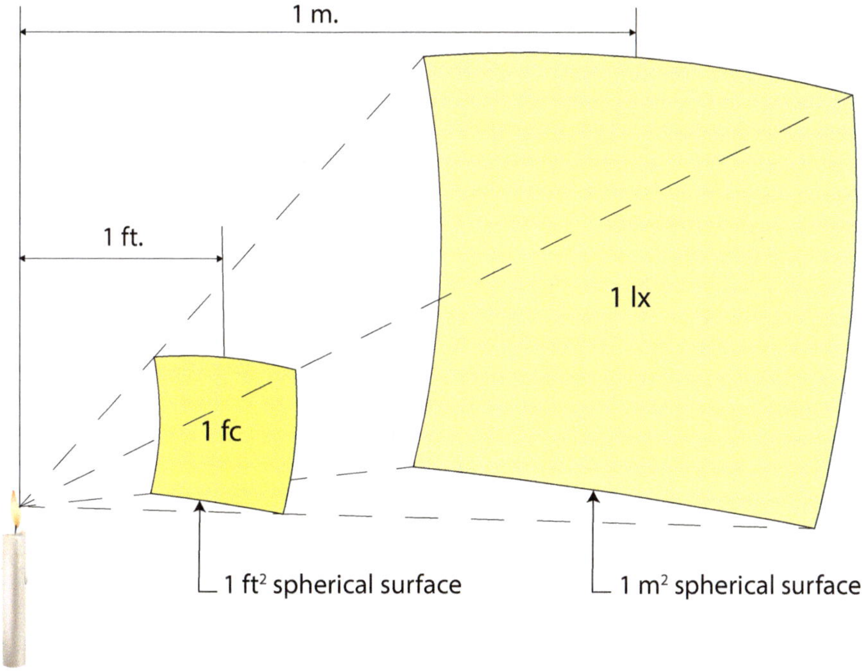

Figure 8.1 Footcandle and lux.

Note that because these values are rounded down to two decimal places the conversions back and forth do not produce the exact same results. However, they are within a few percent, which is accurate enough for a lighting practitioner's needs.

These two are often rounded off to simply conversion so that:

$$1\text{fc} = 10\text{ lx and } 1\text{ lx} = 0.1\text{ fc}.$$

8.2.3 Reflectance

The unit used to measure reflectance is the same as that used for a luminous surface because both are surfaces giving off light. Whether that light originates inside the surface or outside the surface and bounces off, they are effectively the same thing, so the units of measurement are cd/ft^2 and cd/m^2.

The reflectance of permanent surfaces, such as wall paint and ceiling tiles, is an important component of many lighting calculations. The lighting designer requests the relevant information from the architect and/or interior designer or, if they don't know, finds it. For example, an architect may not know the reflectivity of the acoustic ceiling system they are specifying but the lighting designer can contact the manufacturer for that information, which is provided as a percentage of light striking the ceiling.

While the reflective value of permanent surfaces can be determined during the design process, the reflectance of other surfaces that are part of visual tasks, such as tabletops, notepads, materials being assembled, etc. may not be known. Since designers can't know the reflectivity of the surfaces of visual tasks, they focus on the illuminance of those surfaces, not light reflected to the eye, in their calculations.

8.3 Illuminance Recommendations

The units of measurement and calculations of illuminance would be meaningless without context, which gives them relevance. Over the course of a project, lighting designers determine the various activities that will occur in the spaces they are lighting and turn to authoritative guidance or recommendations for appropriate light levels. These form the basis for determining a project's illuminance level, although the lighting designer is free to modify them based on the client's goals, the project's aesthetics, the designer's experience, and other factors. Where does this guidance come from?

In North America the most authoritative recommendations come from the Illuminating Engineering Society (IES), which publishes a series of recommended practice (RP) documents covering a wide range of project types. Each RP includes a discussion of common visual tasks for that project type and provides recommended illuminance ranges.

Some manufacturers also publish illuminance guides on their websites.

8.3.1 Modifying illuminance recommendations

Generalized illuminance recommendations are just that – recommendations based on a combination of studies, historically provided light levels, and input from experienced lighting practitioners. However, there are several reasons a lighting designer would choose to modify those recommendations.

8.3.1.1 Age of occupants

To accomplish the same visual task with equal ease, the average 70-year-old requires twice as much light as the average 20-year-old. For spaces where the average occupant is over 65 years old, the IES suggests that recommended light levels should be doubled (IES Light and Design Lighting Committee, 2020).

8.3.1.2 Owner's or user's desires

In discussions with the owner or users of a space, the designer may discover that there is a preference or desire for light levels that are higher or lower than the recommendation. Provided those light levels don't impair task performance or exceed energy conservation power allowances, most lighting designers would be expected to follow the owner's or users' direction.

8.3.1.3 Design approach

Most illuminance recommendations are for average illuminance across the plane where tasks are performed. However, there are design approaches or strategies that are more area specific. In open offices, for example, one strategy is to provide lower ambient illumination levels while ensuring that supplemental task lighting on desktops meets the recommended (or owner/user's desired) illuminance.

8.3.1.4 Alternative recommendations

There are alternative guidelines that permit modifying the recommended illuminance. For instance, *IES TM-24-13 An Optional Method for Adjusting the Recommended Illuminance for Visually Demanding Tasks Within IES Illuminance Categories P through Y Based on Light Source Spectrum* (IES Visual Effects of Lamp Spectral Distribution Committee, 2013) permits designers to reduce the illumination levels within certain illuminance ranges and for certain applications if the spectrum meets certain conditions. It may be worth noting that this document has not been updated since its publication in 2013 and is not an ANSI standard.

8.3.1.5 Other methods of improving visual performance

In some instances, increasing illumination levels is either not desirable or not possible. However, there are other methods for improving task visibility beyond simply providing more light (Howard et al., 2022). They include:

- increasing the size of the detail in the task, perhaps by using a magnifying device;
- increasing the contrast of the task;
- changing the color of the task components to increase color difference;
- selecting a light source with more appropriate color rendering properties;
- if the visual target is moving, reducing its speed;
- if the visual target is presented for a limited time, increasing the presentation time.

8.4 Illuminance Calculations

Some lighting practitioners begin to lay out the lighting plan knowing that when the layout is finished the luminaire intensity can be selected to achieve their illuminance goal. Other practitioners establish the illuminance goal first, and work toward meeting that goal as they lay out luminaires. The two approaches work equally well. Once a practitioner has reviewed illuminance recommendations, considered modifying those recommendations to suit the specifics of a project, arrived at an illuminance goal or target, and completed a preliminary luminaire layout, that layout can be tested through illuminance calculations and modified if needed. Depending on the information and level of detail the designer needs, one or two hand calculations and/or specialized software may be used.

8.4.1 Point method

The simplest calculation is the point, or point-to-point, method. As its name implies, this method calculates the amount of light leaving a single luminaire and landing on a single point on a surface. The basis of the point method

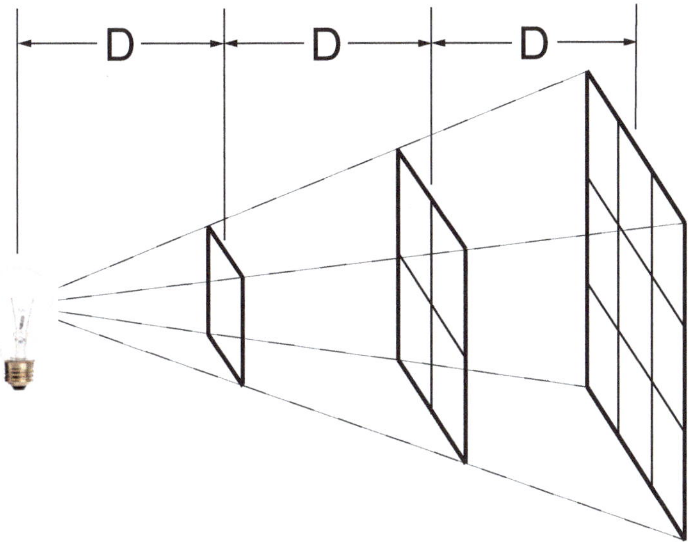

Figure 8.2 Inverse square law.

is the inverse square law which states that doubling the distance between a source and the target causes the intensity to drop to one-fourth the original intensity, as shown in Figure 8.2.

There is one caveat associated with the point method, which is that the light source must be small enough to behave as a point. In this case, that means the largest diameter or diagonal of the source or luminaire must be no more than 1/5 of the distance between the source and the target. This is called "the 1/5 rule." This means that a 4" (100 mm) downlight, for example, must be at least 20" (500 mm) from the target surface, which it typically would be.

As another example, a 2'×2' (600×600 mm) square downlight has a diagonal measurement of 2'-10" (850 mm), so it would have to be 14'-2" (4.25 m) above the target surface. These types of luminaires typically produce very wide beams and installing them in such a high ceiling would deliver very little light to the target, so they're rarely used in ceilings of that height. This illustrates a second rule-of-thumb for the point method – it is probably inappropriate for luminaires with an aperture greater than 12" (300 mm).

The basic point method formula is given below eqn. (8.1):

$$E = \frac{I}{D^2}, \tag{8.1}$$

where:
E = Intensity of light (illuminance) falling on the surface. Note that the calculation will provide a result in fc if the distance is measured in feet, or lx if the distance is measured in meters.
I = Intensity of the lamp or luminaire (luminance) in the direction of interest, measured in cp or cd.
D = Distance between the lamp or luminaire and the point of calculation, measured in feet or meters.

D is simply the distance between the lamp or luminaire and the target. Depending on the source of light, though, I may be found in one of several places.

If the light source is a replaceable, directional lamp the performance is determined, measured, and reported by the lamp manufacturer. The assumption made by lamp manufacturers is that the only item of interest is the brightness at the center of the beam, so that is often the only measurement they report. To find I the lighting designer would look at the website or specification document for the lamp to find the center beam cp or cd. Note that different manufacturers use different language such as center beam candlepower (CBCP), middle beam candlepower (MBCP), luminous intensity, etc. to describe this measurement. The key is to look for a measurement given in cd or cp.

For example, the web page for an LED retrofit PAR38 might show "Luminous Intensity 6,000 cd." If that lamp were installed in a track light and pointed straight down onto a surface 10' (3.05 m) below the calculation would be:

$$E = \frac{6,000}{10^2}$$

or

$$E = \frac{6,000}{3.05^2}.$$

These two calculations result in 60 fc or 645 lx at the center of the beam.

If the light source is controlled by a luminaire the performance is determined, measured, and reported by the luminaire manufacturer. Luminaire manufacturers typically provide much more detailed information that includes the luminance at angles other than center beam and, especially if the beam isn't a symmetrical cone, at several angles through the beam. Increasingly, manufacturers are providing this information in a data file, but not printing it on their cut sheets. This is unfortunate because an initial evaluation of a luminaire need not be a full computer calculation. If a manufacturer provides photometric information on the cut sheet, the designer looks for a Candelas table and/or a candlepower distribution graph, as shown in Figure 8.3.

In the lighting industry, a common calculation is for a ray of light that travels straight down to a target surface. Therefore, straight down is usually described as having a vertical angle of 0°, which is also called nadir. In addition, because this is such a common calculation, a ray of light striking a surface at 90° is also described as being "normal" to the surface.

To find the brightness at the center of the beam of a downlight the designer would look at the Candelas table and use the value at the vertical angle of 0°. The candlepower distribution graph is a polar coordinate graph of the values in the Candelas table, so it too can be read to find the intensity at 0°.

Using the information in Figure 8.3, the calculation for the light delivered to a surface 8' (2.44 m) below this downlight would be:

$$E = \frac{860}{8^2}$$

Candelas

Vertical Angle	Candelas
0	860
5	888
10	907
15	929
20	953
25	959
30	919
35	856
40	668
45	23
50	0
55	0
60	0
65	0
70	0
75	0
80	0
85	0
90	

Candlepower Distribution

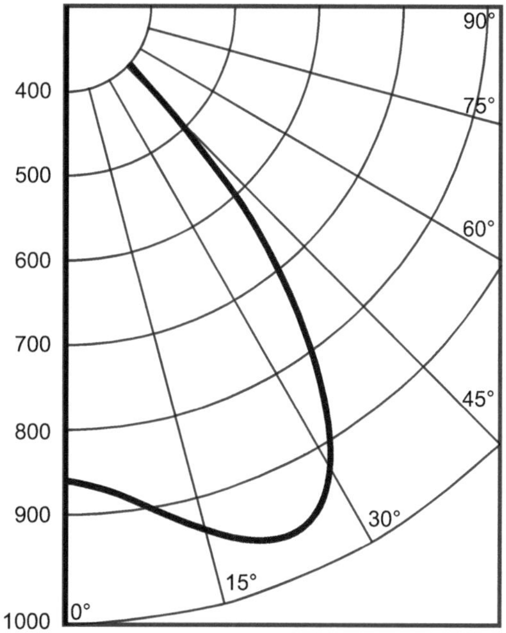

Figure 8.3 Candelas table and candlepower distribution graph.

or

$$E = \frac{860}{2.44^2}.$$

These two calculations result in 13.4 fc and 144.5 lx at the center of the beam.

One common use for the point method is calculating the illuminance from a track light to a target such as a painting on a wall. In this application the light isn't striking the painting at a right angle, so the light is not evenly distributed. To account for this a correction factor is added to the formula. The correction factor is the cosine of the angle of incidence measured from a line normal to the surface, as shown in Figure 8.4.

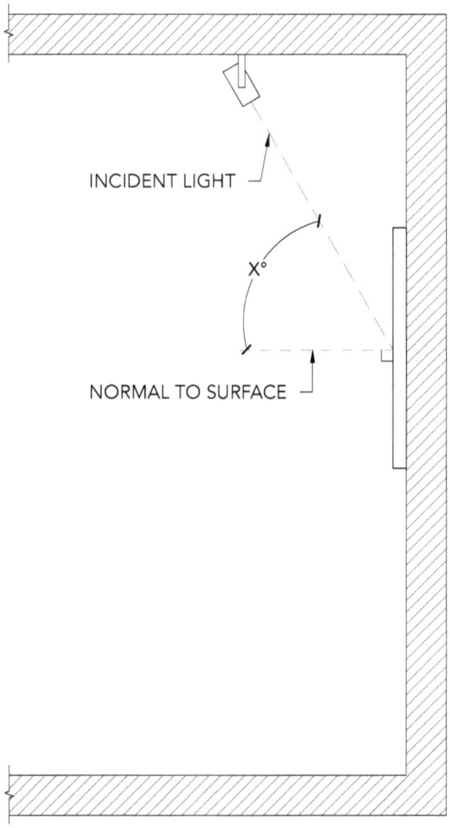

Figure 8.4 Measuring angle of incidence.

The modified formula is eqn. (8.2):

$$E = \frac{I\,(\cos X)}{D^2}, \tag{8.2}$$

where:
E = Illuminance at calculation point, in fc or lx.
I = Intensity of the lamp or luminaire, in cp or cd.
D = Distance between the lamp or luminaire and the calculation point, in ft or m.
X = Angle between surface normal and angle of incident light.

Note that this formula applies to light striking a surface in any orientation: horizontal, vertical, or sloped.

For example, if the 6000 lumen lamp used earlier was applied to Figure 8.4, and if the angle between normal and the center of the beam was 30° the calculation would be:

$$E = \frac{6000\,(\cos 30)}{10^2}$$

or

$$E = \frac{6000\,(\cos 30)}{3.05^2}.$$

These two calculations result in 52 fc and 558.6 lx at the center of the beam.

In this example, it doesn't matter if the target surface is vertical, horizontal, or tilted. As long as the angle of incidence is measured from normal, and the angle of incidence, throw distance, and center beam candlepower are the same, the calculations yield the same results.

8.4.1.1 Accent lighting

As noted earlier, the most common use of the point method is to calculate the brightness of accent lighting. How much accent light is required? The intensity of light is determined by the amount of attention the designer wants to draw to the illuminated target. Chapter 2 explained that our perception of brightness is relative, not absolute. Therefore, the intensity of accent light is also relative – in this case as a multiple of the average illuminance in the space. Table 8.1 provides illuminance ratios of accent light illuminance to average room illuminance, as well as descriptions.

Table 8.1 Accent illuminance ratios.
(based on *ANSI/IES RP-10-20 Recommended Practice Lighting Common Applications*)

Attraction level	Role in visual environment	Target reflectance	Illuminance ratio[1]	Application examples
Subtle	Visual relief	≥50%	1:1	Artwork in offices and residences
		<50%	1:1	
Soft	Visual edge	≥50%	2:1	Featured artwork in offices and residences
		<50%	5:1	
Moderate	Feature	≥50%	5:1	Retail: feature displays Hospitality: destinations such as front desk, featured wall material or artwork
		<50%	10:1	
Strong	Dramatic	≥ 50%	10:1	Retail: exclusive jewellery Worship: reverent focal element
		20:1	40:1	
	Dominant	≥50%	20:1	Retail: exclusive displays or vignettes Hospitality: Exclusive wall materials
		<50%	40:1	

[1]Illuminance ratios are average vertical illuminance at target to average horizontal illuminance at task plane.

8.4.2 Lumen method

Using the point method lighting practitioners can calculate the illuminance at a point. If the goal is to calculate the average illuminance across the room, using the point method would require far too many separate calculations. Instead, practitioners use the lumen method, which is more complicated but also provides more information. One of the advantages of using the lumen method is that it can include correction factors to account for a variety of issues that cause the lighting system to deliver less light over time. These correction factors are used to calculate the maintained illuminance. One drawback to the lumen method is that it can only calculate the average on a horizontal plane, so it can't be used to calculate the illuminance on a wall from wall washers, for example.

The lumen method begins by calculating the average number of lumens per square foot or square meter.

$$E = \frac{L \times N}{A}, \tag{8.3}$$

where:
E = Average illuminance in fc or lx
L = Lumen output of each luminaire
N = Number of luminaires
A = Area of the space in square feet or square meters. As with the point method, if the area is calculated in square feet the illuminance will be in fc and if the area is calculated in square meters the illuminance will be in lx.

Equation (8.3) calculates lumens per area, but that isn't illuminance. The next step is to add a calculation of the distance that the light travels from the luminaires to the plane of calculation (usually desktop height, but other planes can be used). This step includes other factors such as the reflective value of the ceiling and walls, the efficiency of the luminaire, the room geometry, and more in a single number called the coefficient of utilization (CU).

8.4.2.1 Coefficient of utilization

The coefficient of utilization (CU) is specific to each luminaire. Some manufacturers include this information in their cut sheets. Others require practitioner to download a data file and examine it with specialized software. An example CU table is shown in Figure 8.5. To read this table and find the correct CU a practitioner needs information about the room, as follows.

The top row is the ceiling reflectance, measured as a percentage of incident light that is reflected. If the ceiling is acoustic tile the designer can consult the manufacturer's website or documentation for that ceiling system. If the ceiling is painted gypsum board, the designer can consult the paint manufacturer's website or documentation for that color.

The second row is the wall reflectance. If the wall is painted the designer can, once again, consult the paint manufacturer for the reflectance of the specific color. If a wall covering is used, the designer will usually need to obtain a sample and measure its reflectance because very few wall covering manufacturers measure and report reflectance. If the wall is another material, such as wood or used brick, some average reflectance values are available in publications by the IES and others.

Ceiling Reflectance	80%				70%				50%		
Wall Reflectance	70%	50%	30%	10%	70%	50%	30%	10%	50%	30%	10%
Room Cavity Ratio											
0	77	77	77	77	76	76	76	76	72	72	72
1	74	72	71	70	73	71	70	68	68	67	66
2	71	68	66	64	70	67	65	63	65	63	62
3	68	64	61	59	66	63	60	58	61	59	57
4	65	60	57	54	63	59	56	54	58	55	53
5	61	56	53	50	60	56	53	50	55	52	50
6	59	53	49	47	58	53	49	47	52	49	46
7	55	50	46	43	55	49	46	43	46	43	40
8	53	47	43	40	52	46	43	40	46	42	40
9	50	44	40	37	49	43	40	37	43	37	35
10	47	41	37	35	46	41	37	35	40	37	35

Figure 8.5 Example coefficient of utilization table.
Note that all the coefficients are given to two decimal places. Some CUs are higher than 1, but in this example all of them are less than 1 even though the decimal point isn't shown.

The lumen method assumes that 20% of the light passing below the calculation plane returns to the upper part of the room, so there's no need to collect information about the reflectance of the floor material.

The last piece if information needed is for the column on the left – the room cavity ratio (RCR). The room cavity ratio is the result of a formula that processes the room's dimensions and produces a single number between 0 and 10. For rectangular rooms the formula is eqn. (8.4):

$$RCR = \frac{5h\,(L + W)}{L \times W}. \tag{8.4}$$

For rooms that are not rectangular, the formula is modified as eqn. (8.5):

$$RCR = \frac{2.5h\,(perimeter\ length)}{A}, \tag{8.5}$$

where:
h = room cavity height
L = room length
W = room width
A = room area

Note that the formulas return the same RCR values whether the numbers are in feet or meters.

Upper case *L*, *W*, and *A* are the room's full length, width, and area measurements. The lower-case *h*, though, may be only a portion of the room's height. As shown in Figure 8.6, a room can be divided into as many as three layers or "cavities." If the calculation is made for a plane above the floor, such as desktop height, the distance between the floor and the calculation plane is the floor cavity height. If the luminaires are pendant, the distance between the ceiling and a plane through the center of the luminaires is the ceiling cavity height. The remaining height is the room cavity.

Having calculated the room cavity ratio, the practitioner can return to the coefficient of utilization table and read down, from the ceiling reflectance to the wall reflectance to the RCR to find the luminaire's CU. Once found, that value is added to the lumen method formula, as follows eqn. (8.6):

$$E = \frac{L \times N \times CU}{A}. \tag{8.6}$$

Note that painting the walls and/or ceiling a lighter color will increase reflectance which increases the CU. This can result in providing the same illuminance level with lower connected watts, saving energy, and lowering

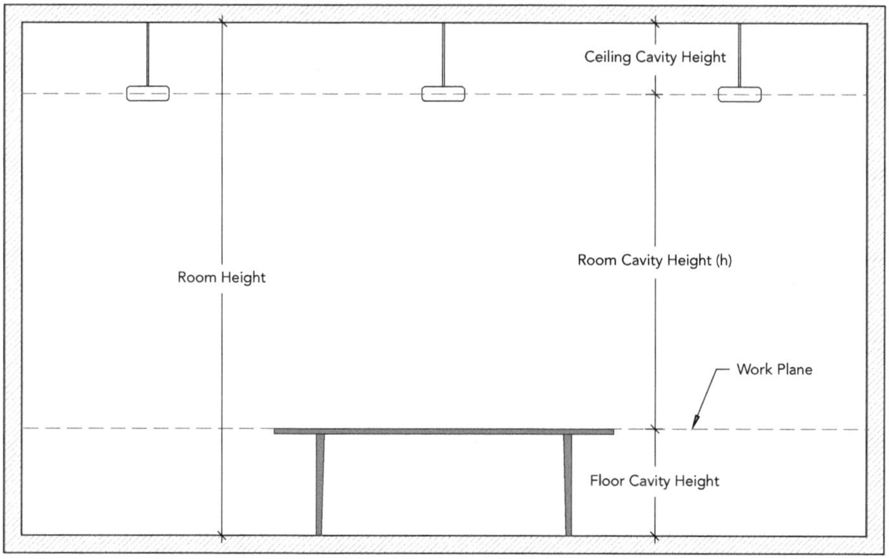

Figure 8.6 Room cavities.

the power bill. Repainting together with LED relighting and the right lighting controls can be a very successful energy savings strategy.

8.4.2.2 Light loss factors

At this point the lumen method is calculating E in fc or lx. However, it's calculating the illuminance on the first day of operation, not the long-term, average maintained illuminance. Calculating the maintained illuminance requires several multipliers, called light loss factors, that account for light lost over time due to changes in the lighting system and the room (Livingston, 2022, IES Maintenance Committee, 2020). The light loss factors are divided into two categories: recoverable light loss factors and non-recoverable light loss factors.

8.4.2.2.1 Recoverable light loss factors

Recoverable light loss factors cause a gradual loss of light on the calculation plane, but there are actions that can be taken to recover, or at least mitigate, the loss. Recoverable light loss factors are:

Lamp lumen depreciation factor (LLD). All lamps emit less light as they age. LLD is a percentage of the lamp's initial lumens calculated by dividing the initial lumens by the mean lumens over the lamp's life. For traditional or legacy light sources, this information should be available from the manufacturer. Table 8.2 gives typical LLD values (Livingston, 2022). LLD has been considered recoverable because eventually traditional lamps are replaced and the light output returns to the original level. That remains true for LED retrofit lamps. Although LED modules in many new luminaires are not replaceable, the industry hasn't moved LLD into the non-recoverable category.

Table 8.2 Typical lamp lumen depreciation values.

Light source technology	Typical lamp lumen depreciation
Standard incandescent	.92
Halogen	.95
T5 Linear fluorescent	.95–.97
T8 Linear fluorescent	.92–.95
Quartz metal halide	.65–.80
Ceramic metal halide	.80–.90
White light LED	.70–.80 (see sidebar below)

LED technology has advanced rapidly, with new generations of LEDs entering the market in rapid succession, and as of early 2024 the industry still does not have a stable, mature product. The result is that the industry doesn't have quality long-term testing of LEDs that provides a reliable LLD. The most recent recommendations are to use .70 (IES Light Sources Committee, 2020; DiLaura et al., 2011; Assist Recommends, 2005). However, the recommended value is the output at the end of the LED's useful life (abbreviated as L70 and described in Chapter 5). This differs from the way other LLDs are calculated (ratio of mean to initial lumen output) and can have a significant impact on delivered light levels and energy consumption (Royer, 2014). For example, an LLD of .70 typically results in an initial light level that is 43% above the target light level, which certainly seems excessive. An LLD of .80 typically results in an initial light level that is 25% above the target light level and reduces energy consumption by roughly 13% vs an LLD of .70. Finally, an LLD of .90 typically results in an initial light level that is 11% above the target light level and reduces energy consumption by roughly 22% vs an LLD of .70. Unless, or until, a more reliable calculation of LED LLD reaches industry consensus designers must use their professional judgement to balance energy consumption, initial light levels, and light levels that the LED system may provide many years later.

There is another way to address this issue. Illuminance calculations assume that all luminaires are at full brightness. However, since nearly all energy conservation codes require luminaires to be dimmable or to have stepped output, and therefore nearly all luminaires include a dimming driver as the standard, other light levels are readily available. One solution is to employ some version of illuminance management, where the lighting system is periodically recalibrated to maintain the target illuminance. In this approach, the designer disregards the LLD and creates a design that provides an initial light level higher than the illuminance target. While there is no industry standard for this, over lighting by 25% to 50% is a reasonable place to start. While the initial light level at full brightness would be above the target, the control system would be set to deliver the target illuminance. Knowing that the illuminance will gradually decline over time, the lighting maintenance plan would include measuring the illuminance, perhaps once per year, and adjusting the control system to make up for any measured light loss.

Luminaire dirt depreciation (LDD). Dirt accumulates on the surfaces of a luminaire. When it does, the reflector, lens, and other components are less effective and light output is lost. The calculation of LDD is complicated and includes the type of luminaire (open or sealed), the CIE classification (direct, semi-direct, etc.) and the environment (clean, moderate, or dirty).

For most interior spaces with typical filtration on the mechanical system, a value of .95 is appropriate. Luminaire dirt depreciation is recoverable because the luminaire surfaces can be cleaned, restoring the luminaire to its original performance.

Similar to LDD, room surface dirt depreciation was once considered a factor in maintained illuminance. Current IES recommendations disregard room surface dirt depreciation due to inadequate research data (IES Maintenance Committee, 2020).

Lamp burnout factor (LBO). All traditional lamp technologies eventually fail – the lamps burn out. If the burned-out lamp is not quickly replaced the room will experience a loss of illumination until it is. The lamp burnout factor is the percentage of lamps that are permitted to fail and not be replaced. It is recoverable because the failed lamps are eventually replaced. With LED luminaires the LBO is disregarded or calculated as 1, that is to say, it has no impact on the calculation.

Including the recoverable light loss factors in the lumen method results in eqn. (8.7):

$$E = \frac{L \times N \times CU \times LLD \times LDD \times LBO}{A}. \tag{8.7}$$

8.4.2.2.2 Non-recoverable light loss factors

Non-recoverable light loss factors cause a permanent change (usually a loss) in the light delivered to the calculation plane from the first day of operation and cannot be overcome later. They include:

Voltage to luminaire factor (VF). Lamps are tested with regulated power that doesn't vary so the results are consistent. In the real world, however, the voltage may be higher or lower than the nominal voltage, especially for renovations in older buildings. Incandescent lamps have the greatest response to changes in voltage, with a 1% voltage change causing as much as a 3% change in output. LED drivers can compensate for considerable changes in voltage (DiLaura et al., 2011) so that, like lamp burnout factor, it can be disregarded or calculated as having a value of 1 (no impact on the calculation) for LEDs.

Ballast factor (BF). Fluorescent lamp output is calculated based, in part, on the use of a standard reference ballast. Commercial ballasts may drive the lamps in a way that produces more or less light than the reference ballast.

Ballast manufacturers publish the ballast factor for each of their products, so the information is easy to find. Ballast factor is becoming an archaic component of the lumen method because it doesn't apply to LED drivers. It can be disregarded or calculated as having a value of 1 (no impact on the calculation) for LEDs.

Thermal factor (TF). Lamps and luminaires are tested at a regulated temperature to ensure consistent results (generally 72 °F or 22 °C). If they are used at considerably higher or lower temperature the light output is likely to be affected. Thermal factor is not an issue for most interior lighting designs. When in doubt, designers can contact manufacturers, who may have thermal factor data available for their luminaires.

Reflector factor (RF). While considerations of the luminaire's reflector are not a formal part of the lumen method, it must be considered. Many luminaires, such as recessed downlights, are offered with a variety of reflector colors and finishes. The photometric information is typically based on a clear specular reflector, which is the most efficient option. When other reflectors are specified, they will reduce the luminaire output, which should be part of the overall assessment of luminaires, layout, and target light level. Light level reductions, given as decimal numbers or percentages, are usually provided on the specification sheet.

The full lumen method calculation, including the reflector factor, is eqn. (8.8):

$$E = \frac{L \times N \times CU \times LLD \times LDD \times LBO \times VF \times BF \times TF \times RF}{A}. \tag{8.8}$$

Often, all the light loss factors are combined, resulting in eqn. (8.9):

$$E = \frac{L \times N \times CU \times LLF}{A}. \tag{8.9}$$

The formula can be rearranged to determine the number of luminaires (F) eqn. (8.10) required provided that the target illuminance and luminaire output (L x N) are known:

$$F = \frac{E \times A}{L \times N \times CU \times LLF}. \tag{8.10}$$

A worksheet for lumen method calculations is provided on the following page.

Average Maintained Illumination Calculation Worksheet – Lumen Method

1. Room Name: ______________________________

2. Design Illuminance Level: ______________________________

3. Luminaire Manufacturer & Catalog Number: ______________________________

4. Lumens per Luminaire ______________________________

5. Fill in Sketch:

Reflectance

Ceiling ________

Wall ________

Floor ________

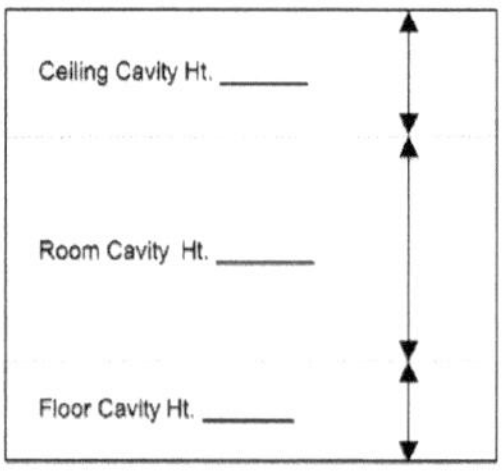

Room Size

L ________

W ________

Area ________

6. Determine Room Cavity Ratio:

$$RCR = \frac{5 \times \text{Room Cavity Height} \times (\text{Length} + \text{Width})}{\text{Length} \times \text{Width}} = \_\_\_\_\_\_\_\_$$

Or

$$RCR = \frac{2.5 \times \text{Room Cavity Height} \times (\text{Room Perimeter})}{\text{Room Area}} = \_\_\_\_\_\_\_\_$$

7. Coefficient of Utilization (CU) from Luminaire Cut Sheet: ________________

8. Determine Light Loss Factors (LLF):

Non-Recoverable		Recoverable	
Voltage Factor	________	LDD	________
Ballast Factor	________	LLD	________
Thermal Factor	________	LDD	________
Reflector Factor	________	RSDD	________

Product of all Light Loss Factors ________

9. Calculate illuminance level achieved:

$$\text{Maintained Illuminance} = \frac{\text{Lumens/Luminaire} \times \text{Luminaires} \times CU \times LLF}{\text{Area}} = \_\_\_\_\_\_\_\_$$

8.4.3 Illuminance calculation software

The point method and lumen method are quick, easy calculations that can be used to determine whether a given luminaire is appropriate. However, when detailed calculations are required practitioners turn to specialized software for a more granular analysis of the distribution of light and the resulting illuminance. Calculations that software can perform, and which can't be done by hand, include illuminance across vertical surfaces, such as walls and bookshelves, and daylight contribution from windows and skylights. The programs vary in the features offered, complexity of operation, and cost. Most software offered by manufacturers and not-for-profit organizations is free or low-cost, while software from for-profit companies can be expensive. The trade-off is that the for-profit companies often provide tech support, while the others may only provide a FAQ list and/or online support community. Commercial software may also have a larger set of features and capabilities. The software is available as a stand-alone package for desktop computers, plug-ins to 3-D CAD software, and as a web-based service. In all three cases the workflow is similar, as are the calculation results.

The first step is to create a 3-D model of the space by using the built-in room builder tool or importing a CAD file or, for plug-ins, building the room in the CAD or BIM software. If daylight is to be part of the analysis, the next step (or the first, depending on the software) is to enter the longitude, latitude, and north/south orientation of the room or building. Next, the room is populated with objects and furnishings, either from the software's library or from imported 3-D files. Then the reflective value or the color (typically in RGB values) is assigned to each surface. Manufacturer provided data files for the luminaires are then downloaded and imported into the calculation. The final step is to determine which planes (such as the work plane) and/or surfaces (such as walls) are of interest and attach a calculation grid to each. Part of setting the grid is to assign spacing of the calculation points on the grid.

The completed calculation may include the illuminance level at each calculation point in plan, elevation, and/or isometric views, gray-scale or color renderings, and in many cases can also be used to create fly-through and time-lapse video files.

8.5 Spacing Criterion

A common question when laying out a design is, "What is the maximum distance between luminaires?" The answer comes from the spacing criterion. The spacing criterion is a rough calculation of how far apart luminaires can be spaced while still providing acceptable illumination uniformity. Spacing criterion only apply to direct distribution luminaires. Spacing criterion is specific to each luminaire, and is typically provided as a single number in the luminaire's photometric report, where it may also be identified as spacing criteria, spacing ratio, spacing, SC (spacing criteria), S/MH (spacing to mounting height), etc.

The spacing criterion, such as 1.2, is multiplied by the vertical distance between the luminaires and the target horizontal plane (typically the work plane) to determine the maximum on-center spacing. For example, if recessed downlights are installed in a ceiling that is 6' (1.8 m) above the work plane and their spacing criterion is 0.8 the luminaires can be placed as much as 4'-9" (1.44 m) apart and still provide acceptably uniform light on the work plane.

There are a few important things to keep in mind with spacing criterion. The first is that it is a ratio, not a fixed number. As the ceiling height increases, so does the luminaire's on-center spacing. The second is that while the spacing criterion can be used to calculate evenly distributed light, it says nothing about the amount of light provided. Average room illuminance is calculated using the lumen method.

8.6 Photometric Reports

Manufacturers increasingly seem to assume that designers only use computer calculations to evaluate luminaires, so they are providing fewer, if any, photometric reports. This is unfortunate because, as this chapter has shown, the point and lumen methods are very useful tools. Fortunately, most calculation software can display a version of photometric reports, so the information is still available though not as easily accessed. Although photometric reports are less common than in the past, it is still important that practitioners can read and understand them because they provide very useful information, some of which is essential for lumen method calculations.

8.7 Photometric Files

Computer calculations require digital data files describing the luminaire's directional and intensity output. There are several file formats whose availability is generally geographically based. Fortunately, most software can read several, or all, of the formats.

In North America the most common format is .ies, which was introduced in 1986. It is defined in *ANSI/IES LM-63 Approved Method: IES*

Standard File Format for the Electronic Transfer of Photometric Data and Related Information. The .ies format only contains directional and intensity information. More recently, the IES published *ANSI/IES TM-33 Technical Memorandum: Standard Format for the Electronic Transfer of Luminaire Optical Data* which defines XML files that contain spectral (color) and other information.

The dominant file format in Europe is EULUMDAT, which was introduced in 1990 and updated in 1998. Other file formats include LTF.

8.8 Photometric Tests

Photometric files are generated by measuring a sample luminaire with a goniophotometer. The device holds the luminaire in a fixed position. A mirror on a motorized arm rotates around the luminaire, reflecting light to a detector. The information captured is translated into the relevant file format for use by illuminance calculation software.

8.9 Measurement Standards

It's important that photometric, lifetime, and other luminaire information be derived from measurements performed according to established standards so luminaires can accurately be compared to one another. The relevant standards set requirements for test conditions, test duration, means of measurement, number of luminaires tested, and more. In many cases the IES has taken the lead in developing these standards. Among those relevant to lighting designers are:

> *ANSI/IES LM-63 IES Standard File Format for the Electronic Transfer of Photometric Data and Related Information*. First introduced in 1986, this is the standard used in creating .ies files.
>
> *ANSI/IES LM-79 Optical and Electrical Measurements of Solid-State Lighting Products*. LM-79 is the standard used by lamp and luminaire manufacturers to test and report on light output, wattage, etc. of complete lighting packages – LED luminaires and retrofit lamps.
>
> *ANSI/IES LM-84 Approved Method: Measuring Optical Radiation Maintenance of LED Lamps, Light Engines, and Luminaires*. LM-84 describes the method for testing complete LED lighting packages – lamps, light engines, and luminaires. This is important

because LED performance is linked to their environment, especially temperature, and the environment inside a luminaire, for example, is unlikely to match the environmental conditions specified in LM-80. It is important to note that many manufacturers do not perform LM-84 testing. Instead, they publish lifetime ratings for the raw LEDs using LM-80 data and TM-21 projections, which may or may not align with the actual luminaire performance.

ANSI/IES TM-33 Technical Memorandum: Standard Format for the Electronic Transfer of Luminaire Optical Data describes an XML formatted file expands on the much older LM-63. Introduced in 2018, it expands on LM-63 by adding data for CCT, color rendering, Duv, and much more. It is a significant improvement that allows, for example, for calculation software to report on a specific CCT from a tuneable white luminaire. Unfortunately, as of early 2024 none of the leading software packages are capable of reading TM-33 files.

The EULUMDAT file format was proposed in 1990 and last updated in 1998. It is less formally specified than LM-63 but common in Europe. The file extension is .ldt.

References

Assist Recommends (2005). *LED Life for General Lighting: Definition of Life*. Lighting Research Center, Troy.

DiLaura, D., Houser, K., Mistrick, R.G., Steffy, G.R. (2011). *The Lighting Handbook, 10th Ed.* New York: Illuminating Engineering Society.

IES Light and Design Lighting Committee (2020). *ANSI/IES RP-10-20 Recommended Practice: Lighting Common Applications*. New York: Illuminating Engineering Society.

IES Light Sources Committee (2020). *ANSI/IES LP-4-20 Lighting Practice: Electric Light Sources – Properties, Selection, and Specification*. New York: Illuminating Engineering Society.

IES Maintenance Committee (2020). *ANSI/IES/NALMCO RP-36-20 Recommended Practice: Lighting Maintenance*. New York: Illuminating Engineering Society.

IES Computer Committee (2019). *ANSI/IES LM-63-19 Approved Method: IES Standard File Format for the Electronic Transfer of Photometric Data and Related Information*. New York: Illuminating Engineering Society.

IES Testing Procedures Committee (2012). ANSI/IES TM-21-21 Technical Memorandum: Projecting Long-Term Luminous, Photon, and Radiant Flux Maintenance of LED Light Sources. New York: Illuminating Engineering Society.

IES Testing Procedures Committee (2021). *ANSI/IES LM-80-21 Approved Method: Measuring Maintenance of Light Output Characteristics of Solid-State Light Sources*. New York: Illuminating Engineering Society.

IES Testing Procedures Committee (2020). *ANSI/IES LM-84-20 Approved Method: Measuring Optical Radiation Maintenance of LED Lamps, Light Engines, and Luminaires*. New York: Illuminating Engineering Society.

IES Visual Effects of Lamp Spectral Distribution Committee (2013). *IES TM-24-13 An Optional Method for Adjusting the Recommended Illuminance for Visually Demanding Tasks Within IES Illuminance Categories P through Y Based on Light Source Spectrum*. New York: Illuminating Engineering Society.

Howard, A., Kelly-Waskett, R., Raynam, P., Thorns, P. (2022). *SLL Code for Lighting*. London: The Society for Light and Lighting.

Designing with Light: The Art, Science, and Practice of Architectural Lighting, 2nd Ed. Hoboken: John Wiley & Sons, Inc.

High Color Rendering Can Enable Better Vision without Requiring More Power, *LEUKOS*, 12:1-2, 27-38, DOI: 10.1080/15502724.2015.1004412

PS-10-18: IES Position on LED Product Lifetime Prediction (2018). Available at: https://www.ies.org/advocacy/position-statements/ps-10-18-ies-position-on-led-product-lifetime-prediction/ [accessed February 9, 2024].

The IESNA Lighting Handbook, 9th Edition. New York, NY.

Lumen Maintenance and Light Loss Factors: Consequences of Current Design Practices for LEDs. *Leukos*, 10:2, 77-86. DOI:10.1080/15502724.2013.855613

9

Electricity and Lighting Controls

Abstract

Controls are a key component of any energy-efficient lighting design and are mandated by many building codes. This chapter begins with a basic explanation of alternating (AC) and direct (DC) current. It explains and illustrates relays and forward phase and reverse phase (ELV) dimming. Then, occupancy and vacancy sensors are introduced, along with the four technologies they employ (passive infrared, ultrasonsic, microphonics, and microwave sensing). The grouping of luminaires into lighting control zones is also discussed along with basic control schedules and narratives. Finally, lighting control protocols used for control system communications are reviewed.

Keywords: Alternating current, direct current, dimmer, forward phase dimming, reverse phase dimming, ELV dimming, preset, scene, astronomical time clock, wallbox, relay, dimmer, zone, connected load, vacancy sensor, occupancy sensor, passive infrared sensor, PIR, ultrasonic sensor, microphonic sensor, microwave sensor, TRIAC, SCR, photocell, daylight harvesting, control zone, control station, switch, RDM, DALI, PoE, BMS, wireless controls, time scheduling, astronomical time clock, task tuning, white tuning, luminaire-level lighting controls, lumen power density, control schedule, control narrative, control operations schedule, commissioning.

9.1 Introduction

At its most basic, lighting controls are simple on/off switches that give users minimal control over the light in their environment – electric lights on or electric lights off. Lighting controls can do much more and many of their capabilities, such as turning lights off when the room is empty and dimming lights when daylight is entering the room, are required by electrical or energy conservation codes. Controls also increase occupant satisfaction by giving them control over their environment and save owners money by reducing power consumption.

9.2 Electricity Basics

Electricity is the induced movement of electrons through a material that is conducive to such movement, which is called a conductor. The movement of electrons is called the current and can take two forms: Electrons flowing in one direction is called direct current or DC. Electrons oscillating back and forth is called alternating current, or AC. DC is the type of electricity supplied by batteries and solar panels. AC is the type of electricity powering most devices in homes, stores, and offices. Both are illustrated in Figure 9.1.

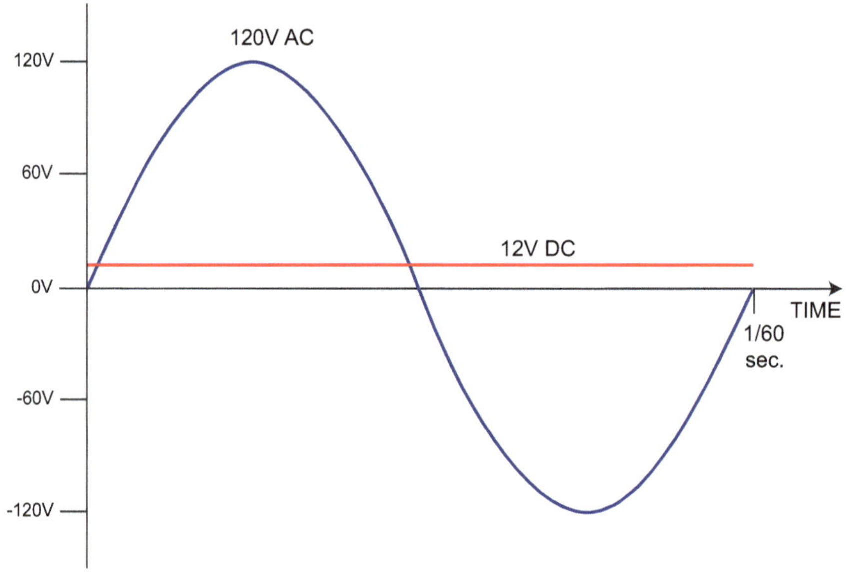

Figure 9.1 Alternating and direct currents.

The force or pressure of the flow of electricity is measured in volts (V). In North America, most electrical devices are designed to operate at a nominal 120 V, although some devices are designed for other voltages. For example, it is more efficient to transmit electricity at higher voltage, so in large buildings it is common to use 277 V for systems such as lighting. In much of the rest of the world 230 V is standard, although voltage varies by country and can also be 110 V, 220 V, or 240 V.

The electric current is of the amount of electricity passing a point and is measured in amperes or amps (A). Lighting practitioners concern with amps usually involves the upper limit that is allowed on a single circuit by a dimmer or relay, or by a protective device such as a fuse or circuit breaker.

Energy is consumed by electric devices (by creating light with an LED, for instance). That consumption is measured in watts (W) and is often referred to as the "connected load" or the "load" on a circuit. Switches and dimmers are also rated by their maximum wattage capacity.

9.3 Importance of Controls

Buildings are the largest consumers of energy worldwide. Building energy consumption doubled between 1971 and 2010 (IEA, 2013) and accounts for 30% of total energy consumption globally (IEA Buildings, 2023). If one wants to reduce energy consumption and the related climate effects, buildings are an obvious target.

Historically, as light has become less expensive people have used more of it. That trend continues. "Electricity consumption for lighting increased in 2022, with greater efficiency not offsetting increased use of lighting. Despite the falling carbon intensity of electricity, CO_2 emissions from lighting rose slightly in 2022" (IEA Lighting, 2023).

Energy conservation codes date back to the 1970s. Early codes sought to reduce consumption by limiting the amount of power that could be used for various purposes. For lighting, the limitation was, and in part continues to be, a maximum connected load per square foot or square meter based on the type of building or the use of the space. Today, most building and space types are limited to less than 1 Watt per square foot or 10 Watts per square meter.

The next step in reducing the power consumption of a lighting system was to require controls that turn lights off when a space is not occupied, and that dim lights (or turn them off) when a space has adequate daylight. These two requirements are a part of every energy conservation code and green program (see Chapter 11). Meeting or exceeding code mandates for lighting controls is now a standard part of the lighting practitioner's scope of work. In addition, technologies such as tuneable white and color changing luminaires require the lighting practitioner to understand the controls needed to operate these systems.

9.4 Benefits of Controls

Lighting control techniques and strategies can provide building owners and occupants with a range of benefits, including:

- Reduced energy consumption. Controls can reduce energy consumption of the lighting system by dimming the lights, using occupancy/vacancy sensors to turn lights off when the room is empty, and using light sensors to dim or turn off lights when adequate daylight is present. The amount of energy that could be saved has varied by study, suggesting that the savings is tied to the control system settings. For example, does the occupancy sensor turn lights off after five minutes or 20 minutes? Nonetheless, a 2012 meta-analysis of 88 published sources showed average savings of 24% for occupancy sensors, 28% for daylight sensors, and 38% for a combined approach (Williams et al., 2012).
- Lumen maintenance. All light sources dim as they age. This can result in spaces that are too bright at the beginning of their life and too dim at the end of their life. Lumen maintenance is a strategy where photosensors or electronics in the luminaires monitor light output and adjust it to hold the brightness steady over time. Lumen maintenance can also be implemented manually by, for example, measuring the room brightness annually and adjusting the control system to make up for any decline.
- Satisfying multi-use requirements. Spaces that have multiple uses, such as conference rooms, require multiple light settings. In the case of a conference room the settings could include in-person meetings, video online meetings, and viewing video presentations. Lighting controls give users access to each setting (also called a preset) at the touch of a button.

9.5 Power Controls

Power to luminaires must be controlled, and electrical circuits must be able to disconnect power, so at a minimum every lighting circuit must be protected by a circuit breaker or fuse. Even when luminaires are not emitting light, they are still consuming power because the driver is in standby mode waiting for instruction. The standby power consumption may exceed what is permitted by the energy conservation code.

In residential lighting, simple switches and dimmers can disconnect the power. In commercial projects, where there are many more circuits and the requirements are more stringent, the default approach is to connect all lighting circuits to a dimmer, or to a relay (a remotely controlled, motorized switch). In many cases dimmers and relays are grouped into a single enclosure called a panel or rack.

9.5.1 Switches

The simplest lighting control device is a switch, which allows users to select one of two conditions: lights on or lights off. Many spaces have more than one layer of light, so they would be expected to have more than one switch. A dining room, for example, might have one switch for the ceiling light(s) and one for the wall sconces.

Most residential energy conservation codes still permit simple on/off switches, but most codes for commercial buildings do not. In commercial spaces a switch must, at a minimum, be paired with a timer or an occupancy sensor to prevent the lights from being left on for long periods of time when the room is empty.

Switches are typically mounted in a metal box that is recessed into the wall, so the switch plate is flat against the wall. They are, therefore, often referred to as wallbox switches. Dimmers used in place of wallbox switches are often referred to as wallbox dimmers (Figure 9.2).

(a) (b) (c)

Figure 9.2 (a) Paddle or "Decora" switch. (b) Toggle switch with sliding dimmer. (c) Rotary dimmer. Courtesy of Lutron Electronics Company, Inc.

9.5.2 Relays

Using relays in a lighting control system separates the intensity control from the power control. The control interface is a low-voltage push button station or a touchscreen. It sends a control signal to the relay, which may not be in the same room or even on the same floor, instructing it to turn power on and off. Relays are typically paired with control signal equipment so that once energized the lighting controls system can dim the luminaire(s). Some relay panels include built-in circuit breakers to protect the electrical system, while others require a secondary circuit breaker panel. A small panel may hold 6 to 8 dimmers and/or relays, while a large panel may hold up to 48, as shown in Figure 9.3. Large projects may require multiple panels per floor.

(a) (b)

Figure 9.3 (a) 4-relay module. (b) 48-relay panel with circuit breakers (not to scale). Courtesy ETC, Inc.

9.5.3 Dimmers

The next step in control sophistication is to replace on/off switches or relays with dimmers, giving users greater control over the lighting. Dimmers reduce the electricity delivered to the luminaire by modifying its sine wave. Two technologies are in widespread use: forward phase dimming and reverse phase dimming. Both types are also called phase cut dimming and two-wire dimming.

9.5.3.1 Forward phase dimming

Forward phase dimmers control the intensity of the lamp by varying the "on" point in each half of the AC cycle (Figure 9.4), to control the amount of power delivered to the lamp. Forward-phase dimmers use either TRIACs (triode alternating current switch) or SCRs (silicon-controlled rectifier). Triac dimmers are the most common wallbox dimmers, while SCRs are typically used in large, centralized dimming systems. Forward phase dimming has been around since the early 1960s and is the most used dimming technology because it is inexpensive and reliable. However, electronic devices don't work well with forward phase dimming, so lighting practitioners should avoid specifying forward phase dimmers to control LED luminaires.

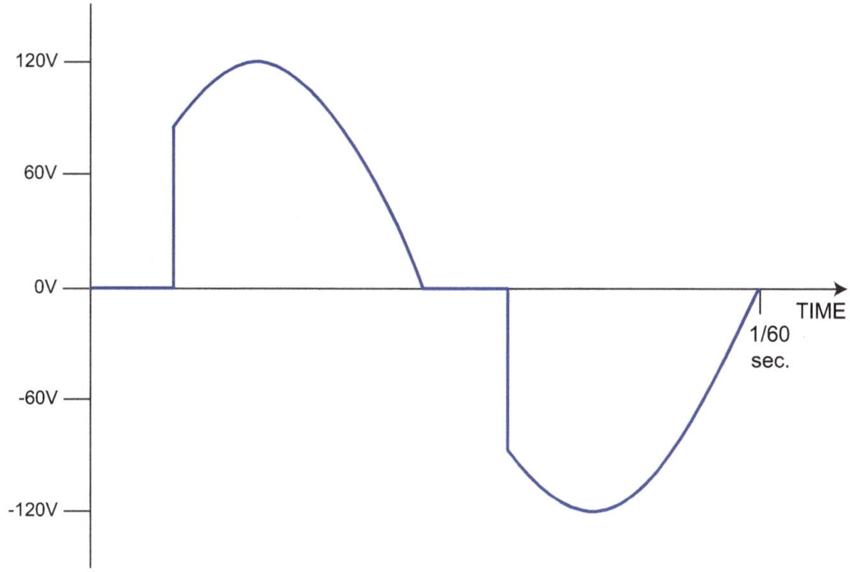

Figure 9.4 Forward Phase Dimming.

9.5.3.2 Reverse phase dimming

Reverse phase dimmers are the opposite of forward phase dimmers. They control lamp intensity by varying the "off" point in each half of the AC cycle (Figure 9.5). They are often referred to as electronic low voltage (ELV) dimmers because they were originally designed to dim electronic transformers for low voltage lamps. Reverse phase dimmers are preferred for dimming the electronic drivers in LED luminaires.

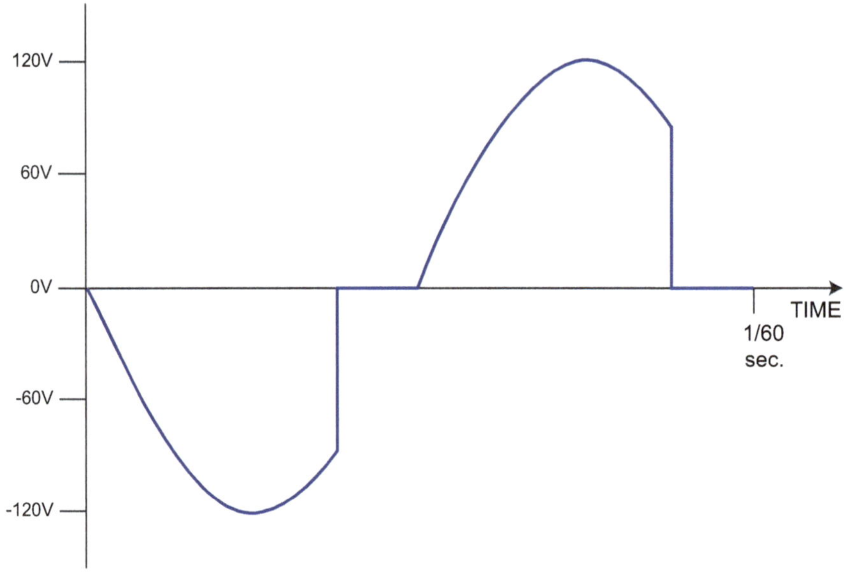

Figure 9.5 Reverse phase (ELV) dimming.

9.5.3.3 Dimming curves

As noted in Chapter 2, the human vision system's response to changes in brightness isn't linear; it is inverse logarithmic (see Figure 9.8). A dimmer that adjusts brightness linearly (emits 70% of full brightness when set to 70%, 10% of full brightness when set to 10%, etc.) will be perceived as a logarithmic change, not a linear change. In this situation, when the dimmer is outputting 70% of full brightness it will be perceived as 90%, and when outputting 10% it will be perceived as 35%. The result is the light will appear to dim quickly at the beginning and more slowly at the end of a fade, which is usually undesirable.

The solution is to match the controller signal output to the driver dimming curve as follows (see Figure 9.6):

- Pairing a logarithmic driver (designed to produce a logarithmic change in output) with a linear controller (designed to produce a linear change in control signal) produces a logarithmic change in brightness that will be perceived as being linear.
- The reverse is also true. Pairing a linear driver (designed to produce a linear change in output) with a logarithmic controller (designed to produce a logarithmic change in control signal), also produces a logarithmic change in brightness that will be perceived as being linear.

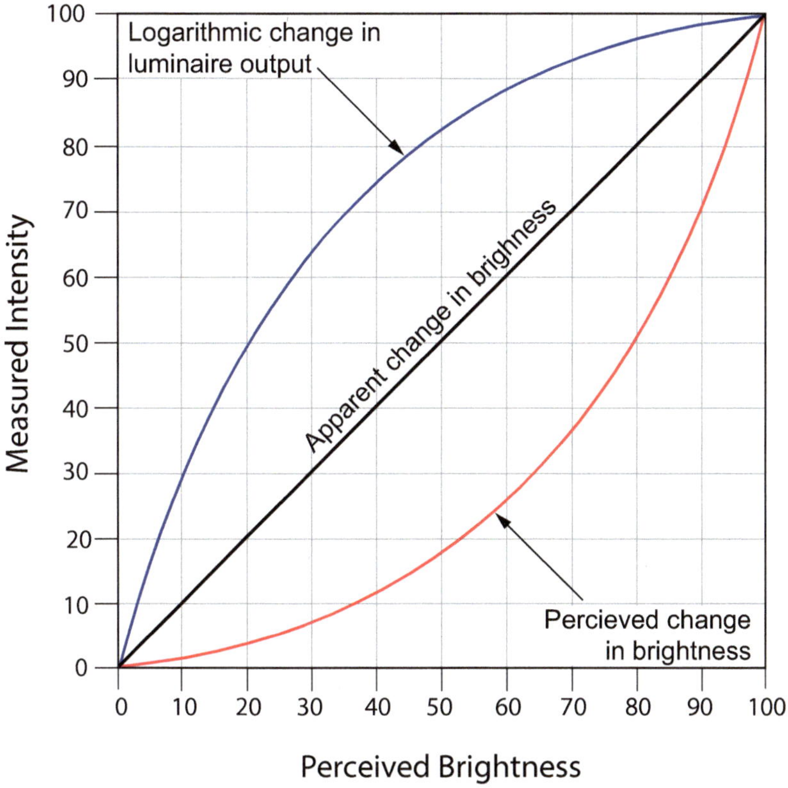

Figure 9.6 Logarithmic dimming output produces linear change in brightness.

9.6 Occupancy and Vacancy Sensors

As noted earlier, commercial energy conservation codes require lights to automatically turn off under certain conditions. All energy conservation codes (see Chapter 10) set requirements for controls in at least some spaces to sense the presence of people and turn off the lights when the space is empty. The underlying idea is that the most energy-efficient luminaire is one that is off. Sensors keep the lights off unless people are present and are more effective than relying on people to remember to turn off the lights when they leave a room. Studies have shown that occupancy sensing reduces energy consumption by up to 50% (Washington State University and the Bonneville Power Administration, 2011).

Sensors have two general configurations and can employ multiple technologies. The two sensor configurations are occupancy and vacancy, and most sensors can be configured in the field, which reduces the number of sensor types manufacturers, lighting practitioners, and contractors need to track on a project. The type of sensor used in a space is dictated by the applicable code.

Sensors can be wired or wireless. Wireless models are powered by long-life batteries (typically with an 8-to-10-year life). Wired models are powered by the network controller or local power packs.

Sensors can be configured as individual devices mounted to a ceiling or wall, wallbox devices that are paired with a manual on/off switch and/or a dimmer (Figure 9.7), or integrated into ceiling-mounted luminaires.

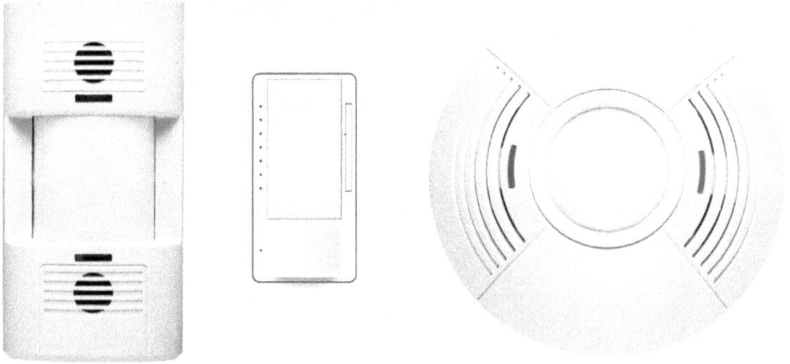

(a) (b) (c)

Figure 9.7 (a) Wall-mounted occupancy sensor. (b) Dimmer switch with built-in occupancy sensor. (c) Ceiling-mounted occupancy sensor (not to scale).

Courtesy of Lutron Electronics Company, Inc.

9.6.1 Occupancy vs. vacancy

Occupancy sensors are configured to turn the lights on when the room becomes occupied, i.e., when it detects people. An internal timer turns the lights off at a specified period after the last detected occupancy. The timer duration may be set at the factory or in the field during commissioning.

Vacancy sensors detect the presence of people, but do not turn on the lights. Vacancy sensors require the lights to be manually activated, but like occupancy sensors an internal timer turns the lights off some time after the last detected occupancy. There are many situations where brief occupancy doesn't require turning on the room's lights, such as when a janitor empties a waste basket, or a colleague enters an office to leave papers on a desk.

By keeping the lights off in these situations, vacancy sensors can save more energy than occupancy sensors.

9.6.2 Occupancy detection technologies

There are four technologies that may be used in occupancy sensors: passive infrared (PIR), ultrasonic, microphonics, and microwave. Each has its strengths and weaknesses. Most manufacturers offer sensors that use two technologies to prevent false activation. In these sensors, both technologies are required to activate the lights but only one is required to keep the lights on. When both technologies detect an empty room, the lights are turned off (IES Lighting Commissioning Committee, 2020).

9.6.2.1 Passive infrared (PIR) detectors

Passive infrared detectors are tuned to the infrared energy emitted by a human body due to its temperature. The advantage is that they respond to changes in heat, not to motion, so they can't be triggered by activity such as drapes billowing next to an open window. There are two disadvantages. The first is that they operate by line of sight, meaning that if a person is behind a barrier, such as a bathroom stall, they cannot be detected. The second is that other sources of heat in the room can trigger the sensor.

The best applications for PIR detectors are spaces with clear lines of sight and areas with high air flow (which does not affect PIR sensors), such as warehouses and aisles.

9.6.2.2 Ultrasonic detectors

Ultrasonic detectors sense motion by sending out high-frequency sound and listening to the sound reflections. A change in the reflection due to moving objects in the room triggers the lights. The advantage is that ultrasonic detectors can see around barriers such as bathroom stalls. The disadvantage is that they can be triggered by other movement or changes in the room, such as drapes billowing by an open window.

The best applications for ultrasonic detectors are spaces where direct line of sight is not possible (such as restrooms), and where air flow is well controlled so as not to falsely trigger the sensor.

9.6.2.3 Microphonic detectors

Microphonic detectors have built-in microphones that listen to the sounds in a room. Their strength is a microchip programmed to distinguish between room sounds, such as someone talking or typing, and ambient and external sounds, such as the HVAC system or a car driving by outside. Their weakness is quiet. Someone sitting still and meditating, for example, might be mistaken for an empty room and the lights would be turned off.

9.6.2.4 Microwave detectors

Microwave detectors emit frequencies in the microwave range and analyse the reflected energy, similar to the operation of ultrasonic detectors. Their advantage and disadvantages are also like those of ultrasonic detectors. One additional disadvantage is that the microwave energy can penetrate glass and drywall barriers. Microwave detectors need to be placed so they are not seeing into a hallway or adjoining room in a way that could cause false activation.

Microwave detectors are the most effective technology for outdoor use.

9.7 Photocells

Photocells sense the ambient light in a space, both daylight and electric light. The measured brightness is internally compared to a preset light level. When the measured brightness exceeds the preset level the photocell sends a signal to the control system or the connected luminaires to dim.

Daylight harvesting is the term that describes the use of photocells to adjust the electric lighting in response to the presence of daylight in the room. A simple daylight harvesting system will raise and lower the connected light fixture or group of fixtures to maintain a predetermined light level as daylight entering the room changes during the day. Photocells can also be connected to control systems that are programmed with more sophisticated responses, such as automatically raising or lowering window blinds or shades.

Various studies have shown that, depending on many factors, using photocells for daylight harvesting can result in energy savings of between 20% and 60% (Integrated Daylighting Systems, 2014).

9.8 Control Zones and Presets or Scenes

In most lighting designs, luminaires are grouped for control by their type (downlight, wall washer, etc.) and their location. Each group is called a control zone or zone, and each zone is controlled separately. Most rooms have more than one zone. For example, the conference room in Figure 9.8 has at least six zones: the luminous ceiling; the downlights on either side of the luminous ceiling; the cove lights; the downlights around the perimeter of

Figure 9.8 A conference room with at least six zones of lighting.
Istock.com/luchunyu.

the room; under cabinet light in the far left corner; vertical lights on either side of the wall-mounted video display.

In this room the simplest control setup would have one dimmer for each zone installed next to the door. There are two possible problems with that setup, in addition to code compliance. The first is that a wall plate with six dimmers may be viewed as large and unattractive. The second is consistency. Users would have to manually set each dimmer every time they used the room, which would likely mean the room looks slightly different every time. The solution is to install a small control system that provides a push-button station next to the door. During commissioning, the lighting designer will adjust the brightness of each zone for each room use (in-person meeting, video meeting, video presentation, etc.) and record them into the control system. Each button, when pressed, triggers its assigned set of light levels, which are called presets or scenes.

Many manufacturers will engrave or silkscreen a custom label for each button, as shown in Figure 9.9(a). Some control stations include controls for the associated zones. Figure 9.9(b) shows such a control station, with the zone controls hidden behind the dark, flip-up panel. Touchscreens (Figure 9.10) are also available from many manufacturers.

(a) (b)

Figure 9.9 (a) Preset control station. (b) Preset control station with dimmer controls (not to scale). Courtesy of Lutron Electronics Company, Inc.

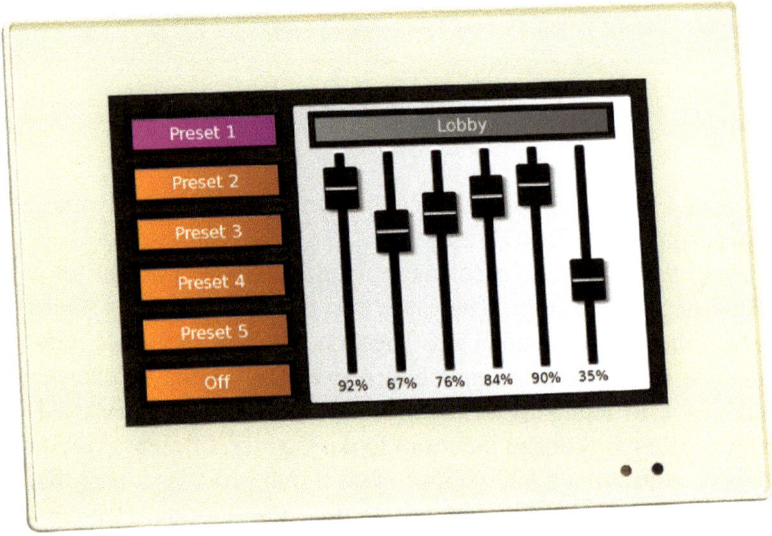

Figure 9.10 Touchscreen controller. Courtesy of ETC, Inc.

9.9 Control Protocols

Dimmers control the light output by modifying and reducing the electricity delivered to the luminaire. This can be problematic with LED luminaires, especially at low light levels, as the electricity delivered is insufficient to properly operate the electronic driver. Flickering is the most common result, but other problems include varying light output across a group of luminaires and some luminaires suddenly turning off.

A more stable, predictable approach is to deliver constant power to the driver along with a control signal that tells the driver how bright it should be. Over time several control protocols have been developed. The protocol (or protocols) used on a project depends on the use requirements, the luminaires, and the control system's capabilities. The most common protocols are described below.

9.9.1 Three-wire control

Three-wire control systems use two wires connected to a forward phase dimmer and a switched third wire supplying line voltage to the driver. The third wire ensures the driver has constant voltage and often allows for a broader dimming range than two-wire dimming. Three-wire control is a patented technology that is not widely licensed, so it is limited to a few manufacturers.

9.9.2 0–10V control

Another method of controlling dimming LED drivers is with a low-voltage direct current (DC) control signal. *ANSI C137.1 0–10V Dimming Interface for LED Drivers, Fluorescent Ballasts, and Controls*, which was published in 2019, is the current standard. With 0–10V control, the LED driver is connected to two pairs of wires. The first pair delivers line voltage. The driver produces a 10 V DC signal and outputs it on the second pair of wires. The control device reduces, or "sinks," the voltage. When the control signal is at 10 V the ballast or driver is producing the maximum designed light output. When the control signal is below 0.5 volts the light source is turned off, and the driver enters standby mode.

C137.1 is based on two earlier standards, ANSI C82.11 Annex A and IEC 60929 which both define 0–10V dimming for fluorescent ballasts. Although these standards technically apply only to fluorescent systems, many existing control systems and LED drivers are still based on these earlier standards, which have an important difference from C137.1. The older standards do not define the minimum output level, leaving that decision to the individual driver or ballast manufacturers. Drivers designed under the old standards dim to a manufacturer-defined level. Lighting practitioners see this noted in the driver selection section of a cut sheet, where drivers are often described as providing

"5% dimming," "1% dimming," or "0.1% dimming." The percentage identifies the minimum output level of the driver. Turning the luminaire off completely typically requires the addition of a switch or relay to disconnect power from the luminaire.

One potential problem with a 0–10V installation is that the standards also permit a 1–10V control signal where the output level below 1 V is not defined, leaving it up to individual manufacturers. Inadvertently mixing 0–10V and 1–10V equipment in a system can lead to several problems and will produce unsatisfactory results.

A second potential problem is the controller's ability to sink the 10 V signal. If too many luminaires are connected to a single controller, it may be unable to fully sink the voltage, resulting in a limited dim range or other performance problems. Unfortunately, very few control manufacturers provide information about this issue on their website or cut sheets. One suggested rule of thumb is to connect no more than 12 luminaires to a single controller. Questions should be directed to the controller manufacturer.

Note that an alternative 0–10V system used in entertainment applications is *ANSI/ESTA E1.3 Entertainment Technology Lighting Control Systems – 0 to 10 V Analog Control Specification.* It defines the controller, not the luminaire's driver, as the source of the 10 V control signal. Devices designed according to this entertainment standard are not compatible with any of the architectural 0–10V standards.

9.9.3 DMX and RDM control

DMX or DMX512 was originally developed in 1986 as an entertainment industry standard for lighting control consoles to communicate with racks of dimmers. It is a robust protocol that has also been adopted by many architectural manufacturers to control multi-colored LED luminaires. The formal standard is *ANSI E1.11 Entertainment Technology, USITT DMX512-A, Asynchronous Serial Digital Data Transmission Standard for Controlling Lighting Equipment and Accessories.*

DMX sends a continuous stream of data for 512 devices or parameters. Each stream of 512 addresses is referred to as a universe. Multiple devices can have the same data address. For example, to have a row of RGB LED cove luminaires behave identically, each luminaire would be given the same address. Most devices require multiple addresses. For example, an RGBW LED luminaire usually requires four addresses to individually control the red, green, blue, and white LEDs.

DMX is unidirectional. The controller sends out the information and the receiving device is expected to follow the instructions. However, because it is one way the controller cannot know when new devices are connected or when a connected device is not responding. In large systems, this can make troubleshooting difficult. The solution is RDM.

ANSI E1.20 Entertainment Technology — RDM Remote Device Management Over DMX512 Networks, or RDM, is a bidirectional extension of DMX. It enables a device connected to the network to broadcast its presence, the parameters that are available to control, and any faults that are detected. It also enables the controller to discover new devices, configure them, and monitor their performance.

9.9.4 Digital addressable lighting interface (DALI)

Digital addressable lighting interface (DALI) is a standard that was introduced in the 1990s as a digital alternative to 0–10V control in architectural lighting. The organization that originally developed DALI was dissolved in 2017. The Digital Illumination Interface Alliance currently certifies DALI products. The original DALI specifications have been revised into DALI-2, which is specified by a series of standards in IEC 62386 (DALI, 2021).

DALI is a four-wire system. Two wires deliver line voltage power, and two additional wires deliver a digital signal. Like DMX/RDM, DALI is a bidirectional system where a unique address can be assigned to every luminaire in a building to control each one individually, or luminaires can share an address. Unlike DMX/RDM information about each luminaire's status in a preset (group, intensity, CCT, etc.) is stored in memory in the luminaire, not in the controller. Luminaires can also be grouped into control zones which can be reconfigured without any adjustments to the wiring. There is no practical upper limit to the number of devices a DALI system can control, although each data loop is limited to 64 luminaires.

9.9.5 Power over Ethernet (PoE) control

Ethernet (technically IEEE 802.3 *Standard for Ethernet*) is a network technology and protocol used to connect computers, printers, etc. Each device sits at one end of a Cat5e or Cat6 network cable, with the other end connected to a network switch or router. Ethernet cables contain four pair of wires but only two of them are used for data. Power over Ethernet, or PoE, is an ethernet extension that allows the switch or router to provide up to 90 W of low voltage DC power over the cable as well.

Using PoE to power and control luminaires essentially makes them computer network devices and presents the possibility of adding a variety of sensors to luminaires. Sensors built into various luminaires within a room could monitor occupancy, daylight, temperature, humidity, carbon dioxide, and more and report conditions to a building management system (BMS). The BMS could be programmed to use temperature data to adjust the HVAC system, for example. As of early 2024, PoE has seen limited adoption in the lighting industry, however it may offer some advantages over traditional installations, including:

- reduced materials and labor installation costs by eliminating conduit and separate power and data wiring to luminaires;
- sustainability benefit of reducing materials used, especially conduit;
- lower installation cost.

9.9.6 Wireless control systems

Wireless controls are broadly available in several configurations and using several communication protocols. Commercial wireless systems differ from residential systems by limiting or eliminating the use of cell phone apps, which can present security and access issues, in favor of dedicated control devices.

Wireless controls offer several advantages over wired controls. They are easy to install in existing buildings since there is no need to open the walls and ceiling to run new control wires. They can also usually be configured wirelessly, decreasing commissioning time. One disadvantage is the potential for electromagnetic interference from high-powered equipment or other wireless devices operating in the same radio frequency band. While manufactures have worked hard to minimize this risk, wireless controls may not be appropriate for mission critical or life-safety related tasks.

9.9.6.1 Wireless devices

Wireless lighting control devices tend to be designed for a single purpose, and therefore have low power requirements. Devices such as wireless occupancy sensors, photocells, and push button scene controllers are available from many manufacturers. Most of these devices use a small, long-lasting battery with a rated life of 8–10 years. Some manufacturers offer battery-free devices that derive their power from the environment through ambient light, change in temperature, or the mechanical energy created by flipping a switch or pushing a button.

9.9.6.2 Infrared (IR) communication

As with PIR occupancy sensors, infrared control devices require line of sight to communicate with one another, which limits their usefulness. In commercial lighting they are most often seen as a hand-held remote control for luminaires, especially for track lighting in high-ceiling museums and galleries.

9.9.6.3 Radio frequency (RF) communication

Most wireless control systems use radio frequency communication, and many of them rapidly jump from one frequency to another to minimize interference. One key consideration with RF devices is the communication protocol because devices using different protocols cannot communicate with one another. There are several widely available protocols, including EnOcean, Zigbee, and Z-Wave, but many manufacturers use a proprietary protocol. When considering wireless lighting control systems, lighting practitioners must consider the devices available from a manufacturer and the protocol used with each option.

9.9.6.4 Bluetooth®

Bluetooth® is a widely adopted point-to-point (single device to single device) wireless protocol. The most common consumer use is wireless audio streaming to speakers, headphones, and in-car systems. Bluetooth Low Energy builds on Bluetooth® by adding the capability to operate in point-to-point, broadcast, and mesh networks. Bluetooth® has introduced Bluetooth Networked Lighting Control, which has been adopted by several dozen controls manufacturers worldwide (Learn About Bluetooth, 2024). In some cases, Bluetooth Low Energy is rebranded by the manufacturer who has licenced the technology making it seem like it is proprietary which can leave lighting practitioners uncertain if devices from two manufacturers can communicate. When in doubt, consult the manufacturer(s) in question.

9.9.6.5 Wireless networks

Wireless devices can be designed to operate on two types of networks: broadcast or fixed, and mesh. In a fixed network every control device communicates directly with a central station, often called a hub, gateway, or repeater. The range of a single hub allows it to connect to the devices in multiple adjoining spaces, but large floors and multi-story buildings require multiple hubs. In some systems the hub is built into a local, wired scene controller with on-board processing and memory. In others, the hub functions like a Wi-Fi router and is wired to a more powerful central controller designed to control a floor, building, or campus of buildings.

In a mesh network every device acts as a hub, and each device is in communication with multiple surrounding devices. The advantage is the multiple paths of a mesh network are less susceptible to inference or disruption. The disadvantage is communications can be slower than with a fixed system and could result in a visible lag between devices or luminaires.

9.10 Control Techniques

Occupancy/vacancy sensing and dimming are hardware-based control techniques. Other techniques, some of which are required by various building and energy codes, tend to be software-based. They include the following:

9.10.1 Time scheduling

Time scheduling is simply setting a consistent schedule for the lighting control system to turn lights on, dim them, and/or turn them off at specific times of day. While scheduling is rigid, it is also very simple to understand, implement, and modify. At its most basic, the clock and time schedule are built into the relevant controller and are programmed manually. The connected lights then follow the same on/dim/off pattern every day. Slightly more sophisticated systems allow the user to set a distinct schedule for each day of the week that reflects the change in usage, typically between weekdays and the weekend.

The most sophisticated controls use an astronomical time clock. During commissioning the astronomical time clock is programmed with the day, date, and time as well as the building's longitude and latitude. The combination allows the software to calculate the time of sunrise and sunset for every day of the year. With this information the lighting control system can provide a wide range of scheduling options, including:

- turn lights on, change intensity, and turn lights off at programmed times of day;
- different schedules for each day of the week;
- modifications to the schedule for specific days, such as keeping the lights off if the business is closed on a holiday;
- scheduling lighting changes relative to sunrise and sunset, such as turning parking lot and facade lights on and off.

9.10.2 Daylight harvesting

Daylight harvesting is the process of using photosensors connected to a lighting control system to monitor the amount of light contributed by windows and skylights. The lighting control system dims the electric lighting in response to the daylight illumination with the goal of holding the illuminance level fairly constant and saving energy by dimming the electric lights when they are not needed. Daylight harvesting is a mandatory component of most energy conservation codes, which define the area around windows and under skylights that are considered "daylit zones," zoning of luminaires in daylit zones, and the control response.

Code required daylight harvesting can be supplemented by having the lighting control system also control the window treatments (drapes, shades, or blinds) to provide control of the daylight entering the space as well as the electric lighting.

Various studies have shown that, depending on many factors, daylight harvesting can save between 20% and 60% (Integrated Daylighting Systems, 2014).

9.10.3 Task tuning

Task tuning is the practice of adjusting the light output and/or spectral power distribution of the luminaires in a space to achieve and maintain the target illuminance and/or spectrum.

Luminaires are specified, in part, by their maximum light output, and most luminaires are available with several output options. It is common practice for lighting practitioners to select a light output that is higher than initially needed so as the LEDs age and dim (i.e., experience lamp lumen depreciation as described in Chapter 8) the lighting control system can be recalibrated to maintain the target illuminance.

A 2018 study showed that task tuning saved an average of 29% of the energy in the buildings studied (Richman & McIntosh, 2018).

9.10.4 White tuning

Due to their small size, LEDs of multiple CCTs or multiple colors can be integrated into a single luminaire. White tuning, or tunable white lighting, is the use of two or more independently controlled LEDs in a single luminaire to enable adjustment of the CCT. White tuning has many applications, and is a common control strategy for implementing circadian lighting (see Chapter 10).

A CCT range of 2200–6500 K is available, although many manufacturers provide a narrower range. The most common means of control is to use two 0–10V signals, one to control intensity and one to control CCT. DMX and DALI drivers are also available from many manufacturers.

9.10.5 Luminaire-level lighting controls

Luminaire-level lighting controls are lighting control systems in which the sensors and controls are integral to the luminaire, allowing each luminaire to respond to the occupancy and/or daylight conditions around it. More importantly, luminaires can be connected, either with wires or wirelessly, to operate as a complete working system that uses factory presets and doesn't require field commissioning (Introduction to Luminaire-Level Lighting Controls, 2022).

Adding a hub may facilitate supplemental or advanced programming and can function as an interface to a larger lighting control system and other building systems.

9.10.6 Multiple control techniques

Unsurprisingly, combining control techniques results in greater energy savings than any single technique alone. This is why energy conservation codes mandate occupancy/vacancy sensing and daylight harvesting. Advanced control designs will add task tuning and other features to increase flexibility and energy savings. Various studies have shown a range of savings, but savings of 60% or greater over a system that only uses on/off switches is a common finding.

9.11 Communicating with Other Building Systems

Advanced lighting controls can communicate with other building systems to send or receive instructions or information. For instance, the lighting control system can raise and lower the window blinds to adjust the amount of daylight entering a space, or the AV system can dim the lights in a conference room when a presentation is about to begin. There are several protocols that various equipment manufacturers may use. When building system interactions are desired the lighting practitioner must coordinate the protocol(s) to be used with other system specifiers and select an appropriate lighting control system (Livingston, 2022). Common communication protocols include:

- RS232 (*TIA-232-F Interface Between Data Terminal Equipment and Data Circuit-Terminating Equipment Employing Serial Binary Data Exchange*) is an older data communications standard used by various types of equipment separated by no more than 50 ft. (15.25 m). It can be used as the communication protocol between two systems, such as an AV system and a lighting system. Such a setup would allow the properly programmed AV system to turn on, dim, and turn off the lighting. Building management systems (BMS) may use RS232 to send instructions to the lighting control system.
- RS485 (*ANSI/TIA/EIA-485-A, Electrical Characteristics of Generators and Receivers for Use in Balanced Digital Multipoint Systems*) connects a series of devices in a daisy-chain arrangement. Each device can communicate with others on the chain. RS485 has a total distance limit of about 4000 ft. (1220 m) and permits up to 32 sending devices and 32 receiving devices.
- UDP (User datagram protocol) was introduced in 1980 and is used to communicate between systems or devices over an ethernet network or the Internet. The formal document is the *Internet Engineering Task Force Request for Comments 768*. UDP is used where speed is critical, and the system can tolerate some loss of data. It is a common AV system to lighting system protocol.

Some lighting control systems can connect to a computer network over ethernet or directly to the Internet, often for commissioning, programming, and maintenance. In many cases the device settings are accessed and configured through a web browser. Such a setup can allow users to control their lighting via their computers or allow technicians to connect to the lighting system for programming, diagnostics, and updating.

9.12 Lighting Controls in Energy Conservation Codes

As discussed throughout this chapter, energy conservation codes seek to reduce energy consumption by, in part, requiring lighting controls that dim lights or turn them off in response to various conditions. The specifics vary by code, but there are many common requirements. Using the *International Energy Conservation Code* (IECC) as an example, the lighting controls requirements include:

- Lighting controls are mandatory. The controls described in the IECC are mandatory for all but a few spaces. Excluded spaces include those that are required to be continuously illuminated for safety or emergency purposes.
- Occupancy sensors are required in the spaces listed in the IECC. In most spaces sensors are only permitted to turn lights on to 50% brightness, and a manual control (i.e., a switch or button) must be activated to turn lights on to full brightness. There is a time limit (20 minutes in the 2018 IECC) for lights to remain on in an empty room.
- Time switch controls (described earlier in this chapter as time scheduling) are required in all spaces that do not have an occupancy sensor.
- Manual light reduction controls are required in most spaces. The controls must permit the light to be reduced by at least 50%.
- Daylight responsive controls to enable daylight harvesting are required in zones adjacent to windows and under skylights and clerestories. Luminaires within these zones must be controlled separately from other luminaires in the room. There is a description for calculating the size and location of the zones.

The second way energy conservation codes limit the energy consumption of the lighting system is by setting a maximum connected load for the lighting system. Typically, there are two options, the building area method and the space-by-space method.

The building area method includes a list of building types (courthouse, library, office, etc.). Each building type has an associated maximum power allowance (called the lumen power density or LPD) measured in watts per square foot or watts per square meter. In nearly all cases, the power allowance is under 1 W/sq. ft. (10 W/sq. m.) The power allowance is multiplied by the floor area of the building to determine the maximum connected load. The lighting design can include any lighting technology, and any number of luminaires, so long as the total wattage does not exceed the maximum allowed.

The space-by-space method is similar, but instead of looking at the building as a whole, it looks at each room or space in the building and assigns it a space type with an associated power allowance. The allowances for each space in the building are added together to arrive at the building's total power allowance for lighting. As with the building area method, the lighting design can include any lighting technology and any number of luminaires, so long as the total wattage does not exceed the maximum allowed.

9.13 Documenting the Control System

As with every other aspect of a building, accurate, detailed, clear lighting control documentation communicates the design intent to the contractor and facilitates construction. The design documentation should include the following:

- The contract drawings will provide information about the hardware used (occupancy sensors, control stations, etc.) and its placement in the building. Drawings may include installation details and one-line or riser drawings that schematically show the flow of power and control signals through the system.
- The contract drawings will also contain a control schedule, which may be organized by room or by control zone number.
- The control device schedule and hardware specification sheets will identify the manufacturer and model of each component (push button station, occupancy sensor, relay panel, etc.) and the options that are selected.
- The control specification will provide additional requirements for the contractor and manufacturers. In the United States, this is often Section 260943 Lighting Controls, using the Construction Specifications Institute's MasterFormat standard.
- The contract drawings or a supplemental document will contain the control narrative and/or a control operations matrix.

9.13.1 Lighting control schedule

A control schedule organizes the lighting system by room and control zone. It lists each zone's purpose, fixture type, fixture quantity, load, and control protocol to the luminaires, if any. Table 9.1 is an example of a control schedule for the conference room in Figure 9.8.

9.13.2 Lighting control narrative

The lighting control narrative is a written explanation of the control system's general operation that is often developed early in the project. During the project's design phase, the control narrative may be expanded to describe the specific operations of devices, such as occupancy sensors, and a space-by-space or room-by-room explanation of the control system's functions. An alternative to expanding the control narrative is to provide a matrix or schedule of the lighting control operations.

Table 9.1 Sample lighting control schedule.

Conference room lighting control schedule							
Zone no.	**Purpose**	**Fixture type**	**Fixture quantity**	**Fixture load**	**Total load**	**Dimmer/ relay type**	**Control protocol**
1	Luminous ceiling panels	L1	8	45 W	360 W	Relay	DMX/RDM
2	Center downlights	L5	16	12 W	192 W	Relay	0–10V
3	Cove	L4	46	20 W	920 W	Relay	0–10V
4	Perimeter downlights	L5	26	12 W	312 W	Relay	0–10V
5	Wall slots	L10	2 × 9 ft.	5 W/ft.	90 W	Relay	0–10V
6	Under cabinet	L12	10 ft.	4.4 W/ft.	44 W	ELV	NA

9.13.3 Lighting control operations schedule

While the control schedule shows the lighting system on a zone-by-zone basis and how each zone is wired and controlled, the control operations schedule shows the functions of the control system on a room-by-room basis, including what control components are used in each room and, where appropriate, how they function, as shown in Table 9.2.

Table 9.2 Sample lighting control operations schedule.

	Occupancy sensor				**Time clock**		**Wall-mounted controls**				**Daylight sensor**		
Space type or name	**Occupancy**	**Vacancy**	**Time out (minutes)**	**Auto on level**	**Scheduled on**	**Scheduled off**	**On/Off**	**Dimmer switch**	**Preset station**	**Touchscreen**	**On/Off**	**Dimming**	**Maintained light level (lux)**
Conference Room	X		20	50%					X			X	350
Private office		X	10					X				X	300
Open office					8 am	7 pm	X					X	500

9.14 Commissioning

To ensure that a lighting control system is operating properly and providing the functionality and energy savings intended, it must be commissioned. Commissioning is the process of confirming that the correct luminaires, lamps, ballasts, drivers, and controls are properly installed, and that the controls are correctly configured, calibrated, and programmed to meet the design intent and the owner's needs. The benefits of commissioning include (IES Lighting Control Systems Committee, 2020):

- verification that the building systems perform as intended;
- reduced energy consumption and operating costs;
- user acceptance and satisfaction;
- enhanced marketability and value of commercial property;
- full accountability by project participants for the quality of their work and reduced risk exposure for the project team.

Commissioning begins during the design process. The lighting practitioner uses the owner's requirements and their own professional experience to select a control system that will satisfy the project's requirements. The requirements are revised and updated as needed throughout the design process, and the final requirements are incorporated into the design documentation.

On small projects, the commission process may be as short as three steps. First, the lighting practitioner will visit the site during construction to confirm that the correct luminaires (with all their components) are correctly installed. Second, once the lighting system is energized the lighting practitioner will work through the system control zone by control zone to confirm that the luminaires are wired correctly. Third, working alongside the lighting controls commissioning technician, each component of the lighting control system will be configured (e.g., setting the timeout on occupancy sensors), scenes are set and stored in memory, and the control system is functioning as intended.

On large projects the owner may hire a third-party commissioning agent who will work with the project team during the design phase to develop a comprehensive plan for commissioning all the building's systems. They will also take the lead role in the commissioning process and may be involved in writing operations manuals and training the building operations staff. As an independent third party, the commissioning agent is not connected to the design team, contractor, or manufacturers. The agent acts in the best interests of the owner to ensure that the building performs as designed and intended, and that performance meets the owner's requirements.

References

Control Protocols Working Group (2010). *ANSI E1.20 Entertainment Technology - RDM Remote Device Management Over DMX512 Networks*. New York: Entertainment Services and Technology Association.

Control Protocols Working Group (2018). *ANSI E1.11 Entertainment Technology USITT DMX512-A Asynchronous Serial Digital Data Transmission Standard for Controlling Lighting Equipment and Accessories*. New York: Entertainment Services and Technology Association.

Control Protocols Working Group (2021). *ANSI E1.3 Entertainment Technology Lighting Control Systems – 0–10 V Analog Control Specification*. New York: Entertainment Services and Technology Association.

DALI (2021). Available at: https://www.dali-alliance.org. [accessed March 24, 2024].

IES Lighting Commissioning Committee (2020). *ANSI/IES LP-8-20 Lighting Practice: The Commissioning Process Applied to Lighting and Control Systems*. New York: Illuminating Engineering Society.

IES Lighting Control Systems Committee (2020). *ANSI/IES LP-6-20 Lighting Practice: Lighting Control Systems – Properties, Selection, and Specification*. New York: Illuminating Engineering Society.

Integrated Daylighting Systems (2014). Available at: https://www.gsa.gov/system/files/GPG_Findings_015-Integrated_Daylighting.pdf. [accessed March 29, 2024].

International Code Council, Inc. (2017). *2018 International Energy Conservation Code*. County Club Hill, IL: ICC Publications.

International Energy Agency (2013). *Modernising Building Energy Codes*. New York: United Nations Development Program.

International Energy Agency (2023). *Buildings*. Available at: https://www.iea.org/energy-system/buildings. [accessed March 22, 2024].

International Energy Agency (2023). *Lighting*. Available at: https://www.iea.org/energy-system/buildings/lighting. [accessed March 22, 2024].

Introduction to Luminaire-Level Lighting Controls (2022). Available at: https://lightingcontrolsassociation.org/2022/06/15/introduction-to-luminaire-level-lighting-controls/. [accessed March 29, 2024].

Learn About Bluetooth (2024). Available at: https://www.bluetooth.com/learn-about-bluetooth/use-cases/lighting-control/. [accessed April 11, 2024].

Livingston, Jason (2022). *Designing with Light: The Art, Science, and Practice of Architectural Lighting, 2nd Ed.* Hoboken: John Wiley & Sons, Inc.

National Electrical Manufacturers Association (2022). *ANSI C137.1-2022 0–10V Dimming Interface for LED Drivers, Fluorescent Ballasts, and Controls*. Rosslyn, VA: National Electrical Manufacturers Association.

Richman, E.E., McIntosh, J.A. (2018). *Advanced Lighting Control System Performance: A Field Evaluation of Five Systems*. Richland, WA: Pacific Northwest National Laboratory.

Washington State University and the Bonneville Power Administration (2011). *Bi-Level Office Lighting with Occupancy Sensors*. Pullman, WA: Washington State University.

What is Power over Ethernet (PoE)? (2024). Available at: https://www.cisco.com/c/en/us/solutions/enterprise-networks/what-is-power-over-ethernet.html. [accessed on March 24, 2024].

Williams, A., Atkinson, B., Garbesi, K., Rubinstein, F. (2012). *Quantifying National Energy Savings Potential of Lighting Controls in Commercial Buildings*. Berkley, CA: Lawrence Berkeley National Laboratory.

10

Light and Health

Abstract

Science has given us a great deal of knowledge of the impact of light on human health. This chapter introduces circadian lighting and the metrics used to measure the impact of light on the circadian entrainment. The chapter goes on to discuss the light-related health issues of UV disinfection, glare, and flicker. It discusses how lighting is a component of building programs focused on occupant health and programs aimed at keeping hazardous substances out of products, including luminaires.

Keywords: Circadian entrainment, circadian lighting, ultraviolet light disinfection, flicker, CS, circadian stimulus, EML, equivalent melanopic lux, melanopic equivalent daylight illuminance, M-EDI, RoHS, Red List, melanopsin, germicidal UV, GUV, ultraviolet C, UV-C, upper room air disinfection, flicker, visible flicker, glare, direct glare, reflected glare, veiling reflection, disability glare, unified glare rating, UGR, WELL building standard.

10.1 Introduction

Light, especially sunshine, has long been associated with health and well-being, even if the mechanism at work wasn't understood. Over time, it has been shown that the benefits of exposure to sunshine include increased vitamin D and boosts to serotonin levels in the brain. Today, we have a deeper understanding of how particular portions of the electromagnetic spectrum, when received through the vision system, support human health in a variety of ways. Other portions of the electromagnetic spectrum have additional, health-related uses.

10.2 The Circadian System

Most people are familiar with the rod and cone cells found in the retina. Rod cells provide black and white vision at very low light levels. Three types of cone cells – long-wavelength sensitive or L-cones, medium-wavelength sensitive or M-cones, and short-wavelength sensitive or S-cones (often misidentified as red, green, and blue cones) – provide color vision at most light levels. The signals they produce result from the breakdown of pigments within the cells when they are exposed to light.

In 2002 a fifth light-sensitive cell was discovered – the intrinsically photosensitive retinal ganglion cell or ipRGC (Berson et al., 2002). The ipRGCs are distributed across the retina and have a peak sensitivity between 459 and 484 nm, which is in the blue region of the visible spectrum. The pigment that provides this sensitivity is called melanopsin. Although their full functionality is still a subject of research, they are known to output non-visual information to the suprachiasmatic nucleus (SCN) in the brain (IES Light and Human Health Committee, 2018). The SCN is the central pacemaker of the circadian timing system (body processes that occur on a 24-hour cycle) and regulates most circadian rhythms in the body (Neuroanatomy, Nucleus Suprachiasmatic, 2023).

10.2.1 Circadian entrainment

While the day is 24 hours long, humans have internal circadian cycles that range from 23.5 to 25 hours. If these internal cycles operated independently from the external day/night cycle, many people would drift in and out of synchronization with the day. Synchronizing circadian rhythms with the day/night cycle is called circadian entrainment and is implicated in many health outcomes. Circadian rhythms influence sleep, hormone release, appetite, digestion, and body temperature regulation. Disrupted circadian rhythms can cause "drowsiness, poor coordination, and difficulty with learning." They also "increase the risks of obesity, diabetes, mood disorders, heart and blood pressure problems, and cancer, and can worsen existing health issues" (Circadian Rhythms, 2023).

Maintaining circadian entrainment is important for human health and the most important contributor is the ipRGCs. Since ipRGCs respond to light, lighting design has the potential to maintain or improve human health by supporting circadian entrainment.

Note that while circadian entrainment and circadian lighting are topics that continue to be addressed by many manufacturers, organizations, publications, and practitioners, a 2023 IES publication, RP-46, states that circadian entrainment isn't an issue in the general population, but lighting plays an important role in supporting other aspects of health (IES Light and Health Committee, 2023). It remains to be seen if the lighting industry has overreacted to the discovery of ipRGCs and their role in human health. Readers are advised to pay attention to industry publications. This chapter assumes that the industry has not over-reacted.

10.2.2 Factors affecting circadian entrainment

At its simplest, circadian lighting focuses on bright days and dark nights. A white light source that provides 400 lux at the eye is likely to be adequate for circadian entrainment (Interview: A New Method for Evaluating Circadian-Effective Lighting, 2023). However, when examined more closely several aspects of light can contribute to circadian entrainment, singly or in concert. They are:

> Spectrum. IpRGCs have greatest sensitivity between 459 and 484 nm, which is in the blue region of the visible spectrum. Early on this was interpreted to mean that high CCT light would support circadian entrainment. However, since CCT is the result of the perception of the entire visible spectrum, it is possible to have high CCT light sources with minimal energy in the ipRGC's sensitivity range. To understand the effectiveness of a light source it is necessary to measure and evaluate the spectral power distribution (SPD), not just the CCT.
>
> Timing. The time of day when the ipRGCs are stimulated affects circadian entrainment. In the lab, scientists have shown that exposure between 6 pm and 6 am can delay the start of the circadian cycle, while exposure between 6 am and 6 pm can advance the cycle (Non-Visual Effects of Light, 2018, IES Light and Human Health Committee, 2018). Both are said to "phase-shift" the circadian system. In practice, it is not possible to tune a single lighting system to every person in a room, so the focus is on encouraging circadian entrainment by delivering stimulation at a consistent time of day, usually in the morning.
>
> Duration. Lower light levels can have the same impact as higher light levels if exposure is longer. Recent research "suggests that the degree of phase shift increases exponentially with... exposure

duration" (IES Light and Human Health Committee, 2018). It also suggests that short exposure, less than an hour or so, is not effective.

Intensity. Early on it was assumed that higher levels of light reaching the retina would have a stronger effect. Subsequent research sometimes confirmed that assumption and sometimes didn't. Currently, illuminance is not considered an appropriate metric for circadian entrainment by the IES (IES Light and Human Health Committee, 2018) although others, such as guidelines for Circadian Stimulus, suggest that SPD and intensity can trade off (Figueiro et al., 2016).

10.2.3 Measuring circadian lighting effectiveness

There are several key differences between measuring light for circadian effectiveness and measuring for visual tasks. The first difference is that measurements are taken at the eye. What's important is how much light is striking the retina, not how much light is illuminating the task. The second difference is that the SPD is the key piece of data because the calculations includes the ipRGC sensitivity. Without the SPD, calculations cannot be made. Metrics that are based on or derived from the melanopic response of the ipRGCs are often described as measuring "melanopic illuminance" and "non-visual responses." The calculation method used is often dictated by the reason for the calculation. For example, points toward WELL certification use EML or M-EDI (they are treated as equivalent) but not CS (all three are discussed in the following sections).

10.2.3.1 Circadian stimulus

The circadian stimulus (CS) metric was developed at the Lighting Research Center (LRC) at Rensselaer Polytechnic Institute. CS is based on the suppression or delay of the body's production of melatonin, often called the sleep hormone, after a one-hour exposure to the light source being measured.

CS is calculated from the SPD, the spectral sensitivity of the circadian system, and the intensity of light entering the eye. The CS calculation results range from 0.1 (the threshold of effectiveness) to 0.7. The LRC's research indicates "that exposure to a CS of 0.3 or greater at the eye, for at least one hour in the early part of the day, is effective for stimulating the circadian system and is associated with better sleep and improved behavior and mood" (Figueiro et al., 2016). The goal for creating a CS effective lighting design is to provide an SPD and intensity (measured vertically at the eye, not on the work surface) that combine to have a CS of 0.3 or higher in the early part of the day. As the day progresses the CS is lowered by reducing intensity and/or changing the SPD.

Circadian stimulus is the metric used in UL's *DG 24480 Design Guideline for Promoting Circadian Entrainment with Light for Day-Active People*. The LRC maintains an online CS calculator on their website.

10.2.3.2 Equivalent melanopic lux

Equivalent melanopic lux (EML) is based on the melanopsin sensitivity of the ipRGCs in the retina. EML is calculated from the SPD, the spectral sensitivity of the ipRGC, and intensity of light entering the eye. The SPD of the light is multiplied by the melanopsin sensitivity (called the melanopic function) and $V(\lambda)$ (photopic luminous efficiency function, see Chapter 3). The results are then compared to arrive at a melanopic/photopic (M/P) ratio. The M/P ratio is multiplied by the measured vertical light level, in lux, at eye level to arrive at the EML.

EML and M-EDI are the two options for measuring circadian lighting effectiveness in the WELL v2 standard. WELL maintains an online EML calculator on their website.

10.2.3.3 Melanopic equivalent daylight illuminance

Melanopic equivalent daylight illuminance (variously abbreviated as m-EDI, MEDI, M-EDI, etc.) is based on the sensitivities of all five photoreceptors in the eye (rods, three cones, and ipRGCs). M-EDI is calculated from the SPD and the five spectral sensitivities. The results are then compared to the same calculation using D65, a CIE model of daylight.

M-EDI is the metric used in CIE 026:2018 CIE *System for Metrology of Optical Radiation for ipRGC-Influenced Responses to Light*. EML and M-EDI are the two options for measuring circadian lighting effectiveness in the WELL v2 standard.

The IES recommends use of M-EDI as "the most accurate and best available" method of assessing an SPD for daytime alertness and performance (IES Light and Health Committee, 2023). The IES website hosts a TM-30 spectral calculator that includes an option to calculate M-EDI for any SPD.

10.3 Ultraviolet Light Disinfection

Ultraviolet (UV) radiation is a known disinfectant. First used to disinfect surfaces in 1877, water in 1910, and air in 1935, it has effectively been used in water purification and laboratory applications for decades (IES Photobiology Committee, 2021). Germicidal UV (GUV) is often the phrase used to describe UV light used to kill harmful microorganisms including bacteria, viruses, and mold. Scientifically speaking, UV doesn't kill. It "inactivates" these organisms by damaging their RNA and DNA, which prevents them from reproducing (Reed, 2010).

Beginning in 2020, the COVID-19 pandemic brought renewed attention to the possibility of UV disinfection in other applications and facilities. Given that the World Health Organization has said that the COVID-19 pandemic is not the last, or even the worst, pandemic to be expected (The coronavirus..., 2020), lighting practitioners should have basic knowledge of UV disinfection both because clients have begun asking about it and because many manufacturers are now including GUV in both general illumination and specialty luminaires.

The ultraviolet spectrum spans wavelengths from 100 nm to 400 nm and is broken into three ranges. Ultraviolet C (UV-C) has wavelengths from 100 nm to 280 nm. It is a component of sunlight but is filtered out by the earth's atmosphere. UV-C wavelengths from about 200 nm to 280 nm are the most effective for disinfection, with peak effectiveness between 265 and 270 nm.

10.3.1 UV disinfection techniques

LEDs are the least efficient method of producing UV-C. The two technologies used for air and surface disinfection are low-pressure mercury, with a spectrum peaking around 254 nm, and excimer lamps (especially krypton-chloride lamps), with a spectrum peaking around 222 nm. The UV disinfection techniques described below do not provide room illumination, of course. However, the equipment may be specified by the lighting practitioner and connected to and controlled by the lighting control system.

The primary technique for air disinfection that lighting specifiers may become involved with is upper room air disinfection. It typically involves mounting specialty luminaires on the walls of a room or as pendants. The luminaires distribute UV-C across the room above 7 ft. (2.1 m) (Figure 10.1). Pendant luminaires may be dedicated to producing UV-C or may provide direct white light and indirect UV. As the air circulates it brings airborne bacteria and viruses into the UV-C zone and inactivates them. Upper room air disinfection should augment building disinfection (often UV-C disinfection inside air ducts), not be used as the primary disinfection method (IES Photobiology Committee, 2021).

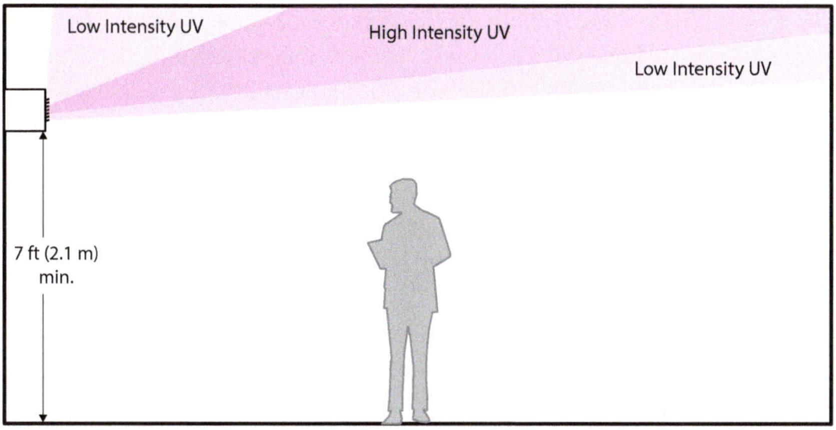

Figure 10.1 Upper room air disinfection with a wall-mounted UV luminaire.

The second technique is internal luminaire air disinfection, although there are not many models on the market. These luminaires have fans to draw air through a vent into an internal chamber where it is exposed to UV-C before being exhausted into the room through another vent. The fan and UV-C can be on a separate circuit from light producing LEDs so that they can disinfect the room even though the luminaire is not producing light. The advantage is that the UV-C is contained in the luminaire, eliminating UV-C exposure to room occupants or surfaces that may be sensitive to UV light. The disadvantages may be fan noise in the room and the need for regular cleaning of the vents and fan.

The third technique uses direct, high intensity UV-C to disinfect both air and surfaces. Due to the potential dangers of direct exposure to UV-C, occupancy sensors and/or other controls are required to if the room is occupied or if someone enters the room. This technique can be integrated into luminaires, especially in healthcare settings. However, it is important that the room be vacated during UV-C treatment, or that anyone using the room wear protective clothing and eye protection. Due to the dangers of direct exposure to UV-C, lighting specifiers should seek additional guidance on its safe use.

10.3.2 UV dangers and safety

While many UV-C luminaires and applications are safe, any application of UV-C must be executed with care to avoid harmful or unwanted effects, including erythema (skin irritation and redness that is similar to sunburn), photokeratitis (temporary damage to the cornea), yellowing of acrylic, fading of textiles, and death of green walls and other plants.

In the United States, the American Conference of Governmental Industrial Hygienists (ACGIH) is the main body that sets exposure limits for UV-C, though other organizations do, too. They do so primarily by establishing threshold limit values (TLVs), which are the level of exposure a worker can be exposed to without adverse effects. The current daily safety limit of 254 nm UV-C for 8 hours is 6 millijoules per square centimeter (mJ/cm^2). These types of measurements are beyond a lighting practitioner's field of expertise, so if there is concern that a luminaire or technique may pose a danger outside consultants may be needed.

Upper room air disinfection is widely accepted as safe because it has a long history of effective use, people do not occupy the UV zone, and most materials are poor reflectors of UV. The biggest danger is to maintenance workers who may be in the UV zone and not realize it (as with workers cleaning air vents). Signage at ceiling vents and access panels or similar precautions should be taken. Provide clearly labeled switches so that the UV-C equipment can be turned off during maintenance.

Room surface disinfection is somewhat less effective than air disinfection. The UV-C light cannot reach areas in that are shaded, and surface dirt can block the UV-C. When used, precautions should be taken to prevent people from entering the room during treatment. Occupancy sensors that cut power to the UV-C source are common. Lockout systems that lock the doors during treatment are also effective. If people must be in the room during treatment, they should wear tightly woven fabrics to protect their skin, and plastic or glass face shields to protect their eyes, and gloves to protect their hands.

10.4 Flicker

Flicker is a term that generally means light with rapid variations in brightness, although there are several types of flicker with distinct scientific names. The three categories of flicker are (IES Vision Science Committee 2, 2020):

> Visible flicker occurs when light varies 3 to 80 cycles per second, or Hertz (Hz). Even small amounts of variation in this range can be visible to some individuals. Visible flicker can trigger photosensitive epilepsy, especially in the 10 to 55 Hz range.
>
> The stroboscopic effect occurs between 80 and 2000 Hz. In this effect, an object appears to move jerkily or jump from one location to another because its motion is not visible during the dips in illumination. This is especially dangerous in industrial applications

because moving machinery can appear to be moving at a different rate, or even appear to be stopped. It can also induce headaches or migraines in susceptible individuals.

The phantom array, or ghost effect, is a change in the perceived shape or layout of objects due to movement. The most common example is a trail of images from a moving car's red LED taillights.

Several factors contribute to visible flicker, including:

- The frequency of the variation, measured in Hz. Higher frequencies equate to less detectable flicker.
- Modulation depth, which is the difference between the brightest and dimmest parts of the flicker measured as a percentage. A lower modulation depth equates to less detectable flicker.
- Motion of the light source or illuminated objects. More motion increases detectable flicker.

Over the years many flicker metrics have been proposed and a half-dozen are in use. Unfortunately, none of them account for all three types of flicker and the many factors affecting flicker perception. Efforts to develop metrics and acceptable values for flicker are ongoing. Readers should consult industry publications for the latest developments.

10.5 Glare

Glare is brightness within the visual field that causes annoyance, discomfort, or overwhelms the visual system. Persistent exposure to glare, such as from a luminaire in an office setting, can cause eye strain, headaches, and reduced employee performance, so it should be considered and avoided as part of the normal lighting design process. Glare occurs when a light source, or a reflection of a light source, is in direct view. One rule of thumb is bright light sources below 45°, measured from a horizontal line through the eyes, are more likely to cause glare than those above the 45°. Well-shielded luminaires have deeply regressed lenses or use baffles or louvers to block direct view of the light source at shallow angles.

10.5.1 Direct glare

Direct glare occurs when the viewer sees an excessively bright light source directly. The light source could be an exposed lamp, the lens of a luminaire, or a view of bright sky or the sun through a window. In any case the source of glare is considerably brighter than the surrounding visual field.

Luminaires designed to limit or prevent glare will limit the angle at which the light source or lens is in direct view. The angle at which the light source becomes visible is called the cut-off angle or shielding angle, depending on how the angle is measured. The shielding angle is measured downward from the ceiling or a horizontal plane at the bottom of the luminaire. The cut-off angle is measured from nadir (0° or straight down from the luminaire). Improving glare control reduces luminaire efficiency because light is absorbed by the shielding materials. Manufacturers strive to find the right balance between efficiency and glare control, generally favoring one or the other in their product line.

10.5.2 Reflected glare

Reflected glare, or a veiling reflection, occurs when the viewer sees a light source or brightly illuminated object reflected from a specular surface such as a glossy magazine or shop window. The visual system tends to focus on the brighter image, so the veiling reflection dominates, and people have difficulty focusing on whatever is behind it.

Reflected glare can be especially problematic in settings where objects are displayed behind glass, such as retail or museum applications, or where glossy materials are viewed, such as polished stone walls in lobbies or glossy magazines on desks. In all cases the object of interest (merchandise for sale, the color and texture of stone, the text on the magazine page) is obscured by the veiling reflection.

10.5.3 Disability glare

The third type of glare, called disability glare, is when the direct view of a light source that makes it difficult to see anything else, as when viewing the headlights of an oncoming car on a dark road. This is not a serious concern in architectural lighting. Although it can occur, as when someone has a view of the sun through a window, it is easily remedied, in this case with drapes or blinds.

10.5.4 Unified Glare Rating (UGR)

Glare is less easily calculated than illuminance (Chapter 8), partly because it is relative not absolute. For example, an exposed lamp seen outside in daylight would be insignificant because the rest of the visual scene is so bright. That same lamp in a dark room would appear to be very bright and would likely be a source of glare. In addition, because of changes in the eye older people are more susceptible to experiencing glare than younger people. Several glare metrics have been proposed over the years, and glare is a subject of ongoing research.

Currently, the best method of assessing glare is the Unified Glare Rating, which is defined in CIE 117-1995 *Discomfort Glare in Interior Lighting*. UGR predicts the likelihood that a direct distribution luminaire will be a source of uncomfortable direct glare. UGR uses a scale with a practical range of 5 to 30. The higher the number the more likely the luminaire will cause discomfort glare. UGR values of less than 10 indicate little to no glare. A good rule of thumb is luminaires should have a UGR of 19 or less.

UGR is the metric used to predict "discomfort glare" in the WELL Building Standard, where points can be earned by using luminaires with a UGR of 16 or lower (Electric Light Glare Control, 2023). UGR is also used by LEED, where a UGR below 19 is an option under the glare control requirement (U.S. Green Building Council, 2024). Most illuminance calculation software can automatically calculate UGR.

10.6 Programs Supporting Occupant Health

While increasing energy efficiency of buildings may be a more widely recognized goal in design, construction, and building management, concerns for the health of tenants and employees is increasingly important to many building owners and employers. Recent studies have found that buildings designed and operated with a focus on occupant health and well-being can charge higher rent (The Financial Impact of Healthy Buildings, 2020), and increase worker productivity (MacNaughton et al., 2015).

Some of the sustainability programs listed in Chapter 11 include requirements that relate to occupant health. One program that is focused on occupant health is the WELL Building Standard. It was introduced in 2014 by the International WELL Building Institute as a performance-based system for measuring, certifying, and monitoring features of the built environment that impact human health and well-being. As of early 2024 the current version was v3.

The idea behind WELL Certification is that employees are a company's most expensive asset, costing more than a building and its operation. Employers who prioritize their employee's satisfaction, health, and well-being are rewarded with gains in productivity and declines in absenteeism and turnover. Buildings are evaluated in seven categories, including one for light that provides illumination designed to minimize disruption to the body's circadian system, enhance productivity, support good sleep quality, and provide appropriate visual acuity where needed. The lighting-related aspects of WELL include:

- Workstations have lighting that provides circadian lighting by meeting specified EML or M-EDI levels.
- All luminaires have a Unified Glare Rating (UGR) of 16 or lower.
- Contrast ratios and illuminance uniformity on the work plane are within specified limits.
- Color rendering meets or exceeds specified values for CRI or TM-30.
- All luminaires are below specified values for flicker.
- Controls provide a specified number of zones per space size and at least three presets to offer a variety of brightness/CCT/distribution options.

10.7 Limiting Hazardous Materials

Luminaires and the materials that go into them can contain substances that are hazardous to humans and/or the environment. The hazard can come from off-gassing (the release of a gas that was dissolved, trapped or absorbed into a material), or the release of substances during use, disposal, or recycling. When hazardous substances are improperly disposed of, they can be released into the air, soil, or water. Several laws and programs address these hazards by prohibiting the use of certain substances in luminaires and other products.

10.7.1 Restriction of Hazardous Substances (RoHS)

As electronic devices have become more prevalent, their disposal has become more problematic because many of them contain one or more substances that are hazardous to the environment and/or public health. The European Commission sought to address the problem in 2006 with the Restriction of Hazardous Substances in Electrical and Electronic Equipment Directive, commonly abbreviated as RoHS. It prohibited the use of 10 substances including heavy metals, flame retardants, and plasticizers, from electric and electronic equipment sold in the European Union, with some exemptions where alternative materials do not exist (Restriction on Hazardous... n.d.). In 2017 the Directive was amended and expanded.

Today many manufacturers produce RoHS compliant luminaires. RoHS compliance is shown on the luminaire's web page and on the specification sheet. RoHS compliance is required for luminaires to receive a CE marking, although many specification sheets that carry the CE mark contain language or marks for RoHS compliance as well.

10.7.2 Red List

The Red List is a component of the International Living Future Institute's Living Building Challenge program. Started in 2006 and updated regularly, the Red List is "a list of chemicals representing the 'worst in class' substances prevalent in the building industry that pose serious risks to human health and the environment." Substances include antimicrobials (when marketed with a health claim), asbestos, formaldehyde, heavy metals, certain volatile organic compounds, and a wide range of chemicals shown to be endocrine disruptors, affect neurological function, and have other adverse health effects (About the Red List, 2022). Luminaire manufacturers producing luminaires that are free of Red List substances may indicate it as part of their Declare label (see Chapter 11) or with a distinct Red List Free label.

References

About the Red List (2022). Available at: https://living-future.org/red-list/. [accessed February 19, 2024].

Berson, David; Dunn, Felice; Takao, Moptoharu (2002). Phototransduction by Reginal Ganglion Cells that Set the Circadian Clock. *Science* 295, 1070-1073. DOI: 10.1126/science.1067262

Brown, Timothy, M. (2020) Melanopic Illuminance Defines the Magnitude of Human Circadian Light Response Under a Wide Range of Conditions. Journal of Pineal Research. 69:1. DOI: 10.1111/jpi.12655

Circadian Rhythms (2023). Available at: https://www.nigms.nih.gov/education/fact-sheets/Pages/circadian-rhythms.aspx#:~:text=Circadian%20rhythms%20are%20the%20physical,and%20temperature%20also%20affect%20them. [accessed April 3, 2024].

Electric Light Glare Control (2023). Available at: https://v2.wellcertified.com/en/wellv2/light/feature/4. [accessed April 7, 2024)

Figueiro, M., Gonzales, K., Pedler, D. (2016) Designing with Circadian Stimulous. Lighting Design + Application, October 30-34.

IES Photobiology Committee (2021). *ANSI/IES RP-44-21 Recommended Practice: Ultraviolet Germicidal Irradiation (UVGI)*. New York: Illuminating Engineering Society.

IES Light and Human Health Committee (2018). *IES TM-18-18 Light and Human Health: An Overview of the Impact of Optical Radiation on Visual, Circadian, Neuroendocrine, and Neurobehavioral Responses.* New York: Illuminating Engineering Society

IES Light and Health Committee (2023). ANSI/IES RP-46-23 Recommended Practice: Supporting the Physiological and Behavioral Effects of Lighting in Interior Daytime Environments. New York: Illuminating Engineering Society.

IES Vision Science Committee (2020). *ANSI/IES LS-7-20 Lighting Science: Vision – Eye and Brain.* New York: Illuminating Engineering Society.

IES Vision Science Committee (2020). *ANSI/IES LS-8-20 Lighting Science: Vision – Perceptions and Performance.* New York: Illuminating Engineering Society.

Interview: A New Method for Evaluating Circadian-Effective Lighting (2023). Available at: https://www.lightnowblog.com/2023/08/interview-a-new-method-for-evaluating-circadian-effective-lighting/. [accessed April 3, 2024].

Neuroanatomy, Nucleus Suprachiasmatic (2023) Available at: https://www.ncbi.nlm.nih.gov/books/NBK546664/#:~:text=Introduction,tracts%20project%20to%20the%20SCN. [accessed April 3, 2024].

MacNaughton, P.; Pegues, J.; Satish, U.; Santanam, S.; Spengler, J.; Allen, J. (2015). Economic, Environmental and Health Implications of Enhanced Ventilation in Office Buildings. Int. J. Environ. Res. Public Health 2015, 12, 14709-14722. https://doi.org/10.3390/ijerph121114709.

Non-Visual Effects of Light: How to Use Light to Promote Circadian Entrainment and Elicit Alertness (2018). Available at: https://www.ncbi.nlm.nih.gov/pmc/articles/PMC6221201/. [accessed April 3, 2024].

Reed, Nicholas (2010). *The history of ultraviolet germicidal irradiation for air disinfection.* Public Health Rep. 2010 Jan-Feb;125(1):15-27. doi:10.1177/003335491012500105.

Restriction of Hazardous Substances in Electrical and Electronic Equipment (RoHS) (n.d.). Available at: https://environment.ec.europa.eu/topics/waste-and-recycling/rohs-directive_en. [accessed April 5, 2024].

The Coronavirus Pandemic is "Not Necessarily the Big One," Senior W.H.O. Official Says (2020). Available at: https://www.nytimes.com/2020/12/29/world/who-covid-pandemic-big.html. [accessed April 2, 2024].

The Financial Impact of Healthy Buildings (2020). Available at: https://realestateinnovationlab.mit.edu/research_article/the-financial-impact-of-healthy-buildings/. [accessed February 18, 2024].

U.S. Green Building Council (2024). *LEED v4.1 Building Design and Construction*. Washington, DC: U.S. Green Building Council

11

Lighting and Sustainability

Abstract

Sustainability is an important consideration in new construction and renovations. This chapter begins by introducing the concept of sustainability programs for buildings and presents information on some of the most widely used sustainability programs for both new and existing buildings. It then discusses product sustainability in the form of take back programs and product labeling for hazardous contents, including ingredient health disclosures and environmental product declarations. It explains the uses of qualified product lists and identifies some of the largest and most commonly used.

Keywords: Sustainability, LEED, Energy Star, Green Globes, qualified product list, environmental product declaration, Designlights Consortium, life cycle assessment, product ingredient list, BREEM, CRREM, take back programs, Declare, UL 2809, Fitwel, DGNB, Green Star.

11.1 Introduction

Building codes are laws that set the minimum standards for a building. Codes are sometimes said to outline the worst allowable building, and most buildings exceed code requirements. Sustainability programs go beyond codes by setting higher and additional requirements on a project and focus on reducing environmental harm, protecting the environment, and/or restoring the environment. Program compliance is voluntary and is usually initiated by building owners who see economic and/or environmental benefits in having their building recognized by a sustainability program. For example, a 2022 analysis of 20,000 U.S. office buildings found that rents in LEED and Energy Star certified buildings were up to 31% higher than similar buildings that were not certified (Green is Good, 2022).

At the beginning of a project the owner should make it clear to the design team that a certain program is to be followed and a certain level of achievement is expected or desired. Introducing these requirements, which are much more stringent than codes, later in the design process can trigger considerable, and expensive, re-design work.

11.2 New Building Sustainability Programs

Sustainability programs focus on minimizing environmental harm caused by building construction and operation. They do so by setting requirements for issues such as site selection, material selection, construction waste management, etc. Many of these programs also have versions that apply to existing buildings. Several of the world's leading green programs are describe below.

11.2.1 LEED

LEED stands for Leadership in Energy and Environmental Design and is probably the most used green building rating system in the world. Founded in 1993, it has evolved into a set of programs for specific project types, including new building construction, interior renovations and fit-outs, homes, and neighborhood developments (LEED Rating System).

To achieve LEED status projects are required to meet certain prerequisites, which vary by project type. Every project type, however, uses the same nine categories for assessment.

Integrative Process – Provide early analysis of building systems and their interrelationships.

Sustainable sites – Limit the impact of the building by, for example, managing storm water runoff, encouraging non-toxic pest control, and using landscaping that reduces the need for irrigation.

Water efficiency – Install systems that reduce water consumption and treat water in an efficient and environmentally sound manner.

Indoor environmental quality – Remove, diminish, and control sources of indoor air pollution within a building and provide user operable controls to guarantee comfortable temperatures.

Materials and resources – Use building materials that have a minimal environmental impact, reduce and control waste, and decrease quantity of materials needed.

Energy and atmosphere – Maximize a building's energy efficiency, use renewable and alternative sources of energy, and adhere to ozone protection protocols.

Location and transportation – Choose infill lots to encourage compact development, site buildings near public transportation and within walking distance of amenities like parks and restaurants, provide facilities to encourage bicycle use to minimize automobile use.

Innovation – Use innovative building features and sustainable building practices and strategies.

LEED projects are awarded points based on the prerequisites and achieving additional, optional credits. The points count toward achieving LEED ratings, which are Certified, Silver, Gold, and Platinum. Quality lighting design can earn a project two points by meeting three of the four following strategies:

Glare control. Use luminaires with a luminance of less than 7000 cd/m^2 between 45 and 90 degrees from nadir or achieve a Unified Glare Rating (UGR) of less than 19 using modeling software to evaluate the design.

Color rendering. Use light sources that have a CRI of at least 90 or use light sources that have TM-30 Color Fidelity Index (R_f) of 78 or greater and Gamut Index (R_g) between 97 and 110.

Lighting control. Provide dimmable or multilevel lighting for 90% of all regularly occupied spaces.

Surface reflectivity. For at least 90% of regularly occupied spaces, use interior finishes with a reflectance greater or equal to 80% for ceilings and 55% for walls. If included in the project scope, use furniture finishes with a reflectance greater or equal to 45% for work surfaces and 50% for movable partitions.

Quality lighting design can also contribute to earning Site Sustainability credit by controlling light pollution, and Indoor Environmental Quality credit by meeting minimum requirements for daylighting (U.S. Green Building Council, 2024).

11.2.2 Green Globes

Green Globes is a program of the Green Building Initiative that "is a science-based, three-in-one whole building certification system that evaluates the environmental sustainability, health & wellness, and resilience of all types of commercial real estate." (Green Globes Certification). Green Globes includes programs to evaluate new construction, existing buildings, core and shell buildings, interiors and interior fit-outs, and multi-family buildings.

The Green Globes system has 1000 points available to every project in the categories of project management, site, energy, water efficiency, materials, and indoor environment. There are no prerequisites: the project owner and team determine which points are relevant, meaningful, and achievable. Projects that achieve at least 35% of the 1000 points earn a One Green Globe certification. Achieving more points allows projects to earn up to a Four Green Globe certification. Quality lighting design can earn up to five points by meeting the following:

- At least 75% of the luminaires have an L70 of 35,000 hours.
- All luminaires are RoHS compliant.
- Design documentation includes manufacturers' cut sheets and other performance documentation for all luminaires.

Quality, energy efficient, lighting design also contributes toward energy performance points by, for example, complying with ASHRAE *Standard 90.1 Energy Standard for Buildings Except for Low-Rise Residential Buildings* (Green Building Initiative, 2022).

11.2.3 BREEAM

Building Research Establishment Environmental Assessment Methodology (BREEAM) was established in 1990 and is administered by the Building Research Establishment. It is similar to LEED in its assessment and goals but uses more stringent requirements. The biggest difference between BREEAM and LEED is that BREEAM uses accredited assessors to evaluate the building for certification while LEED relies on the building's design team to report information toward certification. BREEAM is the most used green building rating system in the UK, and has certified buildings in Europe, the Middle East, North and South America (What is BREEAM?).

BREEAM evaluates new construction, renovations, occupied buildings, and communities using the following seven categories.

Net zero carbon. Strive for net zero buildings by minimizing carbon emissions, employing on-site renewables, and giving credits for energy and carbon reduction. To support this category lighting practitioners should pay attention to LED efficacy and luminaire efficiency, and controls to minimize energy use.

Whole life performance. Extend the lifespan of buildings and their materials by evaluating the sustainability of the design, facilitating a circular journey (also called circular economy) for materials, and providing a framework to balance sustainability goals and targets. To support this category lighting practitioners should look at the recycled and recyclable content of luminaires and look for luminaires with long predicted lifetimes and warranties.

Health and social impact. Support the health of building occupants and the surrounding community by assessing ventilation, thermal and visual comfort, air quality, and social impact. To support this category lighting specifiers should pay attention to recommended illuminance levels and luminance ratios and specify luminaires with high quality color rendering.

Circularity and resilience. Support circularity principles by encouraging durable design, use of responsibly sourced materials, reduced water, and energy consumption, and reducing waste. To support this category lighting specifiers should specify luminaires with high recycled and recyclable content, reduced packaging, and recyclable packaging.

Biodiversity. Design buildings that ensure meaningful benefits locally and avoid losses through strategies that protect and restore biodiversity.

Disclosures and reporting. Provide investors, owners, and operators with necessary information, including ESG (environmental, social, and corporate governance) reporting and solutions, and aligning BREEAM with UN Sustainable Development Goals (which can be found at https://sdgs.un.org/goals).

Quality lighting design can contribute to BREEAM rating by providing daylight and daylight harvesting while controlling glare, providing user and task-appropriate light levels that also control glare, providing occupant accessible lighting controls, and limiting the use of exterior lighting (BRE Global, 2021).

11.2.4 Sustainability programs for existing buildings

11.2.4.1 ENERGY STAR

ENERGY STAR is a program of the U.S. Government's Environmental Protection Agency. It is a broad program that encompasses commercial buildings, homes, appliances, luminaires, and lamps. Energy Star uses a benchmarking system for existing commercial buildings in the U.S. and Canada (Energy Star). A free, online calculator, ENERGY STAR Portfolio Manager, helps establish the current performance of the building measured against similar buildings after the user has entered energy use data for at least 12 full consecutive months for all meters and fuel types, as well as water use and waste data. Once the building is benchmarked, the software provides access to more than 100 metrics to track and access the building's improved performance in energy use, water use, waste management, and greenhouse gas emissions.

The ENERGY STAR system uses a 1–100 scoring system. Buildings earning a score of 75 or higher are eligible for ENERGY STAR certification. The application must be verified by a licensed architect or engineer. The licensed professional verifies that all energy use is accounted for accurately, that the building's characteristics have been properly reported, and that the building operates in accordance with industry standards for indoor environmental quality.

Quality, energy efficient, lighting design can contribute to ENERGY STAR certification by being more energy efficient than the reference buildings.

11.2.4.2 CRREM

Carbon Risk Real Estate Monitor (CRREM) seeks to accelerate decarbonization by increasing the energy efficiency of existing buildings. The underlying idea is that climate change is expected to increase the stringency of building codes and other regulations, increasing the cost of carbon and reducing the value of buildings with poor energy performance. Therefore, CRREM's goal is to educate building owners and assist them in designing energy efficiency retrofits to prevent the properties from becoming stranded (unrentable and unsellable) because of their poor energy efficiency (CRREM Risk Assessment Reference Guide, 2023).

The CRREM-Tool (which as of early 2024 covers 44 countries in Europe, Asia-Pacific, and North America) begins by benchmarking a real estate portfolio or an existing building. It then provides strategies for improvement based on the country, type of building(s) and use.

As with ENERGY STAR, quality, energy efficient, lighting design can contribute by being more energy efficient than the reference buildings.

11.3 Product Sustainability

11.3.1 Take-back programs

It has long been recognized in other areas of building design and construction that it is wasteful, expensive, and environmentally harmful to send building and construction materials to landfills. Instead, many manufacturers of products such as ceiling tiles, carpet, and vinyl flooring have "take-back programs" that accept their old products (for example, old ceiling tiles removed during a renovation) and recycle them into new products (for example, new ceiling tiles).

Unfortunately, as of early 2024 there doesn't appear to be a single major luminaire manufacturer with a take-back program. It can be done, however. LED downlights, for example, could be returned to their original manufacturer when a space is renovated or a building is demolished. The manufacturer would disassemble the luminaire discarding the parts that can't be recycled (such as the LEDs) and recycling or remanufacturing the rest. A housing, for example, could be cleaned, the lens and reflector polished, and new LEDs and a new driver installed. The refurbished luminaire would be nearly identical to a newly manufactured luminaire at a fraction of the cost and environmental harm. Any stigma surrounding refurbished luminaires might be overcome by a lower price and the manufacturer could still make a profit (Livingston, 2022).

11.3.2 Sustainability labeling

To combat greenwashing (making a product appear to be more environmentally friendly or less environmentally damaging than it really is) several organizations have developed systems, usually verified by a third party, to evaluate a manufacturer's products and/or claims. The goal is to give specifiers and customers confidence that the product is as green as it claims.

11.3.2.1 Ingredient health disclosures

Ingredient health disclosures inform the consumer about what goes into the products that surround them every day and are like food ingredient disclosures. Consumers should have the right to know what potentially goes into their bodies through dermal contact, inhalation, ingestion, and other routes of exposure. This strongly encourages manufacturers to remove harmful ingredients from their products and the marketplace. Among the ingredient health disclosures used in building design, construction, and furnishing are Cradle to Cradle Certified, Declare, Global Green TAG, and Health Product Declaration.

11.3.2.2 Environmental product declarations

Environmental product declarations (EPDs) are documents that transparently convey the content and source of materials used in a product. The first step is for a manufacturer to conduct a life cycle assessment (LCA) of their product. An LCA assesses the environmental impact associated with all stages of the life of a product, from raw material extraction and processing, to manufacture, distribution, use, and either recycling or disposal. This information is formatted to fit the EPD system or methodology chosen by the manufacturer. The EPD is then verified by a third party to accurately and completely convey the environmental impact of a product. Several organizations offer EPD systems and services.

11.3.2.3 Declare

In the luminaire manufacturing industry, perhaps the most used ingredient health disclosures is Declare, which was developed by the International Living Future Institute and is used by global manufactures such as Acuity Brands, Erco, and Signify, among others. Declare labels list all product ingredients, use a color code system to flag chemicals of concern, and provide information on the product's final assembly location, life expectancy, and end-of-life options. Manufacturers must disclose all intentionally added materials at or above 0.01% of the final product by weight (Declare Basics, 2022). The label also indicates if the product contains any chemicals on the Living Building Challenge Red List, which "represents the 'worst in class' materials, chemicals, and elements known to pose serious risks to human health and the greater ecosystem that are prevalent in the building products industry." (About the Red List, 2022).

Lighting practitioners can search the Declare database (https://declare.living-future.org) for listed luminaires and their Declare labels. Manufacturers also provide Declare labels on their websites. Figure 11.1 shows an example of the Declare Label for a luminaire.

11.3.2.4 UL 2809-1 environmental claim validation procedure

UL 2809 by UL Solutions (formerly Underwriters Laboratories) is a third-party recycled content validation for manufacturers. UL 2809 describes

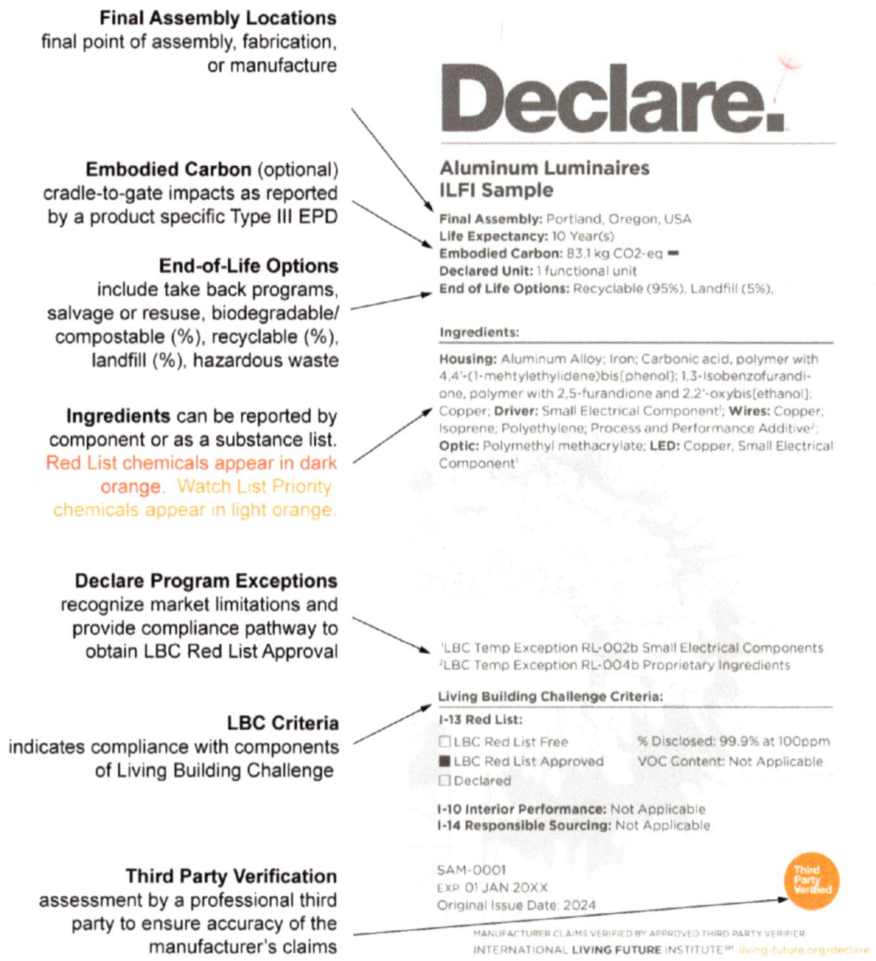

Figure 11.1 Example of a Declare label.
Courtesy of International Living Future Institute

methods for a third party to audit the chain of custody of materials and calculate the final content of manufactured products. The result is a percentage of recycled content that can be displayed on the product, packaging, or marketing materials in a "UL Validated" graphic. To keep their UL 2809 validation manufacturers must resubmit product information annually.

Lighting practitioners can search the UL SPOT database (https://spot.ul.com) for UL 2809 validated luminaires, including those by global manufacturers such as Acuity Brands, Cooper, and Cree. The database also sorts products by a wide range of other rating systems, sustainability credits, and certifications.

11.4 Qualified Product Lists

Another way lighting practitioners can validate a luminaire is through a qualified product list (QPL). QPLs serve a range of purposes. Utilities use them to confirm which luminaires may qualify for a full or partial rebate in new construction and/or renovations. Others use them to validate the performance of a luminaire as suitable for a given program's purpose. Among the available QPLs are:

11.4.1 DesignLights Consortium

DesignLights Consortium (DLC) has the goal of providing data and resources on quality lighting, controls, and integrated building systems to reduce energy, carbon, and light pollution (About the DLC). Manufacturers submit their luminaires to an accredited testing laboratory for evaluation. Luminaires that meet the requirements are added to the QPL, which is often used by electric utilities to qualify a luminaire for a rebate. The DLC maintains four qualified product lists.

> The solid state lighting QPL lists indoor and outdoor luminaires that meet minimum requirements (as of Version 5.1) for light output, efficacy, distribution, controllability, glare, lumen maintenance, spectral quality (i.e., color rendering), electrical performance, warranty, and safety. (DLC Solid-State Lighting SSL Technical Requirements Version 5.1, 2022).
>
> The DLC's LUNA program is designed to mitigate negative impacts of outdoor lighting at night. The LUNA QPL lists outdoor luminaires that meet minimum requirements (as of Version 1.0) for light output, efficacy, distribution, aiming, shielding, BUG rating, controllability, lumen maintenance, spectral quality (i.e., color rendering), electrical performance, warranty, and safety (DLC LUNA Technical Requirements Version 1.0, 2023).
>
> The DLC Networked Lighting Control System program's goal is to speed the adoption of networked lighting control systems for enhanced energy efficiency and centralized management (Networked Lighting Controls). The required capabilities for interior

lighting systems (as of Version 5) are (Networked Lighting Control System Technical Requirements Version NLC5, 2023):

- Networking of luminaires and devices
- Occupancy sensing
- Daylight harvesting/photocell control
- High-end trim (setting a maximum light level)
- Zoning
- Individual addressability
- Continuous dimming
- Cybersecurity

Reported interior lighting control systems (i.e., not required but noted if present) are:

- Control persistence
- Scheduling
- Device monitoring/remote diagnostics
- Type of user interface
- Luminaire level lighting control (LLC, integrated)
- Personal control
- Plug load control
- Emergency lighting
- Color changing/tuning
- Ease of implementation
- Scene control

The DLC also has a program and QPLs for LED horticultural lighting.

11.4.2 ENERGY STAR

ENERGY STAR maintains QPLs for LED retrofit lamps and luminaires. The long recognized retrofit lamp QPL and luminaire QPL were scheduled to be discontinued on December 31, 2024. A new QPL for LED downlights and recessed downlight retrofit kits to replace the luminaire QPL went into effect in November 2023. Downlights must meet minimum requirements for light output, distribution, lumen maintenance, electrical performance, serviceability, thermal performance, controllability, labeling, and warranty (ENERGY STAR Program Requirements for Downlights, 2023).

11.5 Other Sustainability Programs

The programs described in this chapter are among the largest programs in use internationally. However, this list is not exhaustive and there are many programs that are specific to a single country. Other sustainability programs around the world include:

> Fitwel. Fitwel was originally created by the U.S. Centers for Disease Control and Prevention (CDC) and U.S. General Services Administration. Like WELL, Fitwel focuses on the health and wellbeing of the building occupants as well as the surrounding community. Fitwel focuses on location, building access, outdoor spaces, entrances, indoor environment, workspaces, shared spaces, water supply, cafeterias and prepared food areas, vending machines and snack bars, and emergency procedures (Fitwel, 2024).
>
> DGNB. Founded in 2007, DGNB is the acronym for the German Sustainable Building Council. DGNB includes programs for a wide range of new and existing buildings. It focuses on promoting sustainable building practices across Europe. DNG focuses on designing buildings for longevity, conservation of resources, reduction of pollutants and greenhouse gases, and occupants' health, comfort, and well-being (DGNB GmbH, n.d.).
>
> Green Star. Green Star was founded by Green Building Council of Australia in 2003. Developed for the Australian environment but also used elsewhere, Green Star programs promote sustainable fit-outs, buildings, homes, and communities with the goals of reducing the impact of climate change, enhancing health and quality of life, restoring and protecting the planet's biodiversity and ecosystems, resiliency in buildings, fit-outs, and communities, and contributing to market transformation and a sustainable economy (Green Building Council of Australia, 2024).

References

About the DLC (n.d.). Available at: https://www.designlights.org/about-us/. [accessed February 19, 2024].

About the Red List (2022). Available at: https://living-future.org/red-list/. [accessed February 19, 2024].

BRE Global (2021). *BREEAM International New Construction Version 6.0.* Hertfordshire, UK: BRE Global Ltd.

Declare Basics (2022). Available at: https://living-future.org/declare/basics/. [accessed February 19, 2024].

Sustainable Building with the DGNB (n.d.). DGNB. Available at: https://www.dgnb.de/en. [accessed April 11, 2024].

DLC LUNA Technical Requirements Version 1.0 (2023). (Note to Editor: I have a PDF of this guide. There's no proper title page or publication info, so I'm not clear on how to list this. I could use the web page instead.)

DLC Solid-State Lighting (SSL) Technical Requirements Version 5.1 (2022). (Note to Editor: I have a PDF of this guide. There's no proper title page or publication info, so I'm not clear on how to list this. I could use the web page instead.)

Energy Star Commercial Buildings (n.d.). Available at: https://www.energystar.gov/buildings. [accessed February 15, 2024]

ENERGY STAR Program Requirements for Downlights (2023). (Note to Editor: I have a PDF of this guide. There's no proper title page or publication info, so I'm not clear on how to list this. I could use the web page instead.)

Fitwel (2024). Available at: https://www.fitwel.org. [accessed April 11, 2024].

What is Green Star? (2024). Available at: https://new.gbca.org.au/green-star/exploring-green-star/. [accessed April 11, 2024].

Institute fur Immobilienokonomie / Institute for Real Estate Economics, Ulster University, GRESB BV, Universidad de Alicante / University of Alicante, TiasNimbas Business School BV, Tilburg University (2023). *CRREM Risk Assessment Reference Guide*. Austria: Institute for Real Estate Economics. (Note to Editor: I really don't know how to format this one! Help!)

Green Building Initiative (2022). *Green Globes New Construction 2021 (ES + BEQ) Technical Reference Manual*. Portland, OR: Green Building Initiative.

Green is Good: The Enduring Rent Premium of LEED-Certified U.S. Office Buildings (2022). Available at: https://www.cbre.com/insights/viewpoints/green-is-good-the-endurance-of-the-rent-premium-in-leed-certified-us-office-buildings#:~:text=LEED%2Dcertified%20office%20buildings%20tend,for%20non%2DLEED%20certified%20buildings. [accessed February 15, 2024].

LEED Rating System. (n.d.) Available at: https://www.usgbc.org/leed. [accessed February 15, 2024]

Livingston, Jason (2022). *Designing with Light: The Art, Science, and Practice of Architectural Lighting, 2nd Ed.* Hoboken: John Wiley & Sons, Inc.

Networked Lighting Control System Technical Requirements Version NLC5 (2023). New York: DesignLights Consortium. (Note to Editor: I have a PDF of this guide. There's no proper title page or publication info, so I'm not clear on how to list this. I could use the web page instead.)

Networked Lighting Controls (n.d.). Available at: https://www.designlights.org/our-work/networked-lighting-controls. [accessed February 20, 2024].

U.S. Green Building Council (2024). *LEED v4.1 Building Design and Construction.* Washington, DC: U.S. Green Building Council

What is BREEAM? (n.d.). Available at: https://bregroup.com/products/breeam/. [accessed February 18, 2024].

12

Lighting System Financial Analysis and Calculations

Abstract

Especially in renovation and retrofit lighting projects, one consideration is the finances of the new lighting system – what will it cost, how much will it save, and how long will it take for savings to cover the cost? This chapter discusses several methods of analyzing the costs and savings associated with lighting systems. Analysis methods include initial cost, annual operating cost, simple payback, and life cycle cost analysis. Utility bills, legislation related to energy efficiency, utility rebates and incentives, and additional lighting value considerations are all discussed.

Keywords: Initial cost, operating cost, cost of ownership, simple payback, return on investment, rate of return, qualified products list, life cycle cost analysis, delivery charges, supply charges, energy legislation, utility bills, utility rebates.

12.1 Introduction

Discussions about lighting efficiency typically revolve around LED efficacy and luminaire efficiency: how much light is provided for the energy consumed? However, there is another efficiency consideration, which is the efficient use of the owner's money. For example, a luminaire with an efficacy of 100 lumens per Watt costing $500 may be a poor use of money when compared to a luminaire with an efficacy of 90 lumens per Watt costing $400. In this case, the second luminaire provides 10% less light, which may or may not be a meaningful reduction. It also saves 20% per luminaire, which is likely to have an impact on the project's overall cost.

Note that while lighting practitioners weigh the options as described in the previous paragraph, and those options may influence the direction of the design, lighting practitioners are very rarely asked to perform the calculations described in the rest of this chapter. A lighting system is required in every building and space, and architects and owners hire lighting designers because they are looking for benefits beyond lowest cost, such as those discussed in Chapter 1. A professional lighting practitioner is the most qualified person to provide those benefits. However, the fact that lighting practitioners rarely perform these calculations doesn't negate their importance. It only means that if they are made, they are probably made by someone else. The results may, or may not, be shared with the lighting designer as a rationale for examining other design options.

The first few calculations discussed are often called first-level analysis. They are simple but fail to capture the full cost of a lighting system over its entire life. The later calculations are described as second-level analysis. They require more information and involve longer calculation formulas, but more accurately describe the full financial impact of a lighting system.

12.2 Initial Cost

The previous example illustrates the simplest financial analysis – initial cost. An initial cost analysis looks at the cost to purchase (and possibly install) a lighting system by comparing the cost per luminaire and luminaire quantity of Option A to the luminaire cost and luminaire quantity of Option B. Assuming that all other considerations are equal, the lower cost system would be preferred.

However, there may be additional factors to consider when making an initial cost calculation. For example, the local electric utility may offer rebates or other financial offsets for some luminaires, such as those on a qualified products list (see Chapter 11). If the more expensive luminaire qualifies for a rebate, it may offset the higher cost. The calculation for initial cost is given in eqn. (12.1):

$$\begin{aligned}\text{Initial cost} = &((\text{luminaire cost}) \times (\text{luminaire quanity}) \\ &\times (\text{installation cost per luminaire})) - ((\text{luminaire rebate}) \\ &\times (\text{luminaire quanity})). \end{aligned} \tag{12.1}$$

12.3 Annual Operating costs

Initial cost is only a part of the cost of ownership of a lighting system. Other costs include electricity and maintenance which, when factored into the decision, could result in a different outcome compared to the initial cost calculation. Assuming the luminaires of Option A and Option B have similar maintenance requirements, which can then be disregarded as a consideration, annual operating costs are easy to calculate eqn. (12.2).

The owner would inform the lighting specifier of their cost of electricity (typically in kilowatt hours or kWh) or provide a copy of a recent utility bill. The calculation would then be comparing the following result for Option A and Option B. The lighting system with the lower annual operating costs would be preferred.

$$\begin{aligned}\text{Annual operating cost} = {} & \text{electricity cost (per kWh)} \\ & \times \text{annual operation (in hours)} \\ & \times \text{lighting system wattage (in kW)}. \end{aligned} \tag{12.2}$$

12.4 Simple Payback

When considering retrofitting or replacing a lighting system, an owner may ask about the simple payback period, which is the amount of time required to recover the investment in the new lighting system. The simple payback period is calculated by dividing the cost of the new lighting system by the annual energy cost savings (Thumann et al., 2013). The simple payback period, which is typically calculated in years, requires the lighting practitioner to compare the current energy consumption with the energy consumption of the proposed replacement system. After calculating the amount of money saved annually on energy, the simple payback period is:

$$\text{Simple payback} = \frac{\text{initial cost (luminaires, labor, materials, etc.)}}{\text{annual energy savings}}. \tag{12.3}$$

If maintenance costs are a consideration eqn. (12.3) can be modified as follows eqn. (12.4):

$$\text{Simple payback} = \frac{\text{initial costs + annual maintenance cost}}{\text{annual energy savings}}. \tag{12.4}$$

A lighting system that cost \$100,000, for example, and saved \$18,000 per year in energy costs would have a simple payback period of 5.6 years. Table 12.1 is a worksheet to calculate simple payback.

Table 12.1 Simple payback worksheet.

Simple payback worksheet		
Project:		
1	New lighting system equipment cost[1]	$
2	Lighting system installation cost[2]	$
3	Utility rebates and/or tax credits[3]	$
4	Total cost of new lighting system (Line 1 + 2 – 3)	$
5	Existing lighting system wattage (in kW)	
6	New lighting system wattage (in kW)	
7	Kilowatts saved (Line 5 – 6)	
8	Annual operation (in hours)[4]	
9	Kilowatt hours saved (Line 7 × 8)	
10	Electricity cost (per kWh)[5]	$
11	Annual electricity savings (Line 9 × 10)	$
12	Payback period in years (Line 4 ÷ 11)	

Notes:

1. Consult luminaire sales representatives or cost estimator.
2. Consult a cost estimator or contractor.
3. Consult the utility company.
4. Provided by owner.
5. Provided by utility company or found on the owner's utility bill.

An expansion of the simple payback calculation can be used to compare two or more lighting systems. Table 12.2 compares three lighting systems where System A is the least expensive, but Systems B and C are more energy efficient. This calculation simply looks at the simple payback period and shows that the increased cost and increased energy efficiency result in System B requiring 10 years to equal the initial cost and operating cost of System A, while System C would require an additional 7.5 years. These time frames are rather long to wait to see the benefits of higher costs so, based on this analysis, System A would likely be the preferred system.

Table 12.2 Simple payback comparison of three lighting systems.

Lighting system	**Initial cost**	**Cost over System A**	**Operating cost**	**Annual savings vs. System A**	**Simple payback period (yrs.)**
A	$150,000	–	$10,000	–	–
B	$160,000	$10,000	$9,000	$1000	$10,000/$1000=10
C	$165,000	$15,000	$8,000	$2000	$15,000/$2000=7.5

While the simple payback method is easy to use, it should be noted that there are shortcomings and disadvantages. For example:

- It calculates the payback on investment but can't consider additional factors or luminaire options such as increased controllability of the luminaire or superior color rendering.
- It doesn't provide information about the financial impact after the payback period.
- It disregards the time value of money.

12.5 Rate of Return or Return on Investment

Simple payback calculates the amount of time until the financial investment is recovered in savings. The rate of return, or return on investment, is the percentage of investment recovered each year, and is calculated as follows eqn. (12.5):

$$\text{Rate of return} = \left(\frac{\text{annual energy savings}}{\text{initial cost}}\right) \times 100. \qquad (12.5)$$

For example, if a lighting replacement project cost the owner \$45,000 but saved \$15,000 in annual energy costs, the simple payback period, calculated using eqn. (12.3) above, would be 3 years. The rate of return would be:

$$\left(\frac{\$15,000}{\$45,000}\right) \times 100 = 33\%.$$

If maintenance costs are a consideration eqn. (12.5) can be modified as follows eqn. (12.6):

$$\text{Rate of return} = \left(\frac{\text{annual energy savings}}{\text{initial cost} + \text{annual maintenance cost}}\right) \times 100. \quad (12.6)$$

12.6 Life Cycle Cost Analysis

The most common second-level cost analysis of a lighting system is called a life cycle cost analysis. It combines the information used in the previous calculations with additional information about the cost of operating the lighting system to arrive at a comprehensive annual cost.

There are two methods of calculating life cycle costs. The simple method disregards inflation, the cost of interest on loans, etc. and treats all dollars (Euros, Pounds, etc.) as having the same value. The complex method includes the impact of inflation on costs and translates it into the value of today's dollars (Euros, Pounds, etc.). This chapter illustrates the simple method in Table 12.3, which can be contracted for one lighting system or expanded to analyse multiple systems. Calculations related to the complex method can be found in *ANS/IES RP-31* and *Handbook of Energy Audits*, both of which are listed in the References at the end of this chapter.

Table 12.3 Simple life cycle cost analysis worksheet.

Simple life cycle cost analysis worksheet			
	Initial costs	**Lighting System A**	**Lighting System B**
1	Luminaire and control hardware[1]	$	$
2	Installation (labor, materials, etc.)[2]	$	$
3	Total lighting load in kW		
4	Tons of cooling required[3]		
5	Cost of cooling equipment at $___ per ton[4]	$	$
6	Utility rebates	$	$
7	Subtotal initial costs (Lines 1 + 2 + 5 – 6)		
	Annual costs	**Lighting System A**	**Lighting System B**
8	Annual operation (in hours)		Same as A
9	Electricity cost (per kWh)	$	Same as A
10	Annual electricity cost (Lines 3 × 8 × 9)	$	$
11	Electricity for cooling (kW/ton)		
12	Annual cooling costs (Lines 4 × 8 × 9 × 11)	$	$
13	Lamp replacement (if applicable)[5]	$	$
14	Ballast replacement (if applicable)[6]	$	$
15	Driver replacement (if applicable)[6]	$	$
16	Luminaire cleaning (no. luminaires x labor cost/luminiaire)	$	$
17	Total annual cost (Lines 10 + 12 + 13 + 14 + 15 + 16)	$	$

Notes:

1. Consult luminaire sales representatives or cost estimator.
2. Consult lighting controls sales representatives or cost estimator.
3. One ton of cooling is required for every 3.5 kW of lighting load.
4. Consult the project's mechanical engineer or cost estimator for cost of cooling equipment required for the lighting system.

5. Lamp replacement cost calculation depends on the lamp replacement method (spot relamping or group relamping). See Chapter 13. For spot relamping the calculation is eqns. (12.7)–(12.9):

$$\text{Spot lamp replacement cost} = \frac{(\text{cost of one lamp plus labor})\,(\text{no. lamps in system})}{(\text{lamp life})\,(\text{annual operating hours})}. \tag{12.7}$$

Group relamping costs are calculated as:

$$\text{Group relamping cost} = \frac{(\text{no. lamps in system})\,(\text{cost of lamps and labor})}{(\text{lamp life})\,(\text{annual operating hours})\,(\text{percent of life at replacement})}. \tag{12.8}$$

6. The annual cost of ballast and driver replacement is calculated as:

$$\text{Annual ballast or driver cost} = \frac{(\text{ballast or driver cost plus labor cost})\,(\text{no. ballasts or drivers in system})}{(\text{ballast or driver life})\,(\text{annual operating hours})}. \tag{12.9}$$

12.7 Utility Bills

It would be easy if a client's electric bill was simply the number of kWh used multiplied by a fixed rate per kWh. Unfortunately, most utility bills are more complicated than that. The various charges may be the same every month or may be a percentage of other parts of the bill, such as the supply charges. The calculation of some charges is not always clear on a bill, so comparing several bills can help to unravel how the utility arrived at a given amount. The most common components of a utility bill include the following.

12.7.1 Administrative, regulatory, and tax charges

These are monthly charges found on the bill. In some cases, they are a flat amount per month regardless of the amount of energy used, and in others a percentage of the cost of the energy used. They can include things such as a charge for the electricity meter and preparation of the bill, local regulatory charges (sometimes unrelated to the utility but imposed by the local government to fund other government services), and local taxes on the delivery of electricity.

12.7.2 Delivery charges

Delivery charges are the costs for the utility to send electricity to a home or business. They include costs such as power generation, powerline maintenance, and purchase of new equipment. Delivery charges are typically billed at a rate per kW of electricity consumed.

12.7.3 Supply charges

Supply charges are the cost of the electricity the customer has consumed and is typically billed at a set amount per kWh. Some utilities charge a reduced rate beyond a certain amount consumed, so bills may have two amounts related to supply charges.

12.7.4 Calculating utility bills

Given the various charges described the calculation of the cost of energy for a lighting system might proceed as shown in Table 12.4.

Table 12.4 Utility bill calculation.

Monthly administrative, regulatory, and tax charges, pro-rated per kWh	$10.00
Delivery charge, 10 kW @ $4.50/kW	$45.00
Supply charge, first 250 kWh @ $0.15/kWh	$37.50
Supply charge, additional 500 kWh @ $0.10	$50.00
Subtotal	$142.50
Tax, 3%	$4.28
Total bill	$146.78

12.8 Energy Legislation

In some cases, the less expensive lighting system is not a viable option because it consumes too much energy and violates an energy conservation code or other legislation. Energy legislation has been, and continues to be, proposed and passed at every level, from national governments to governments of small villages and towns. In some cases, the legislation is intended to promote certain green technologies such as solar or geothermal. In others the goal is to reduce fossil fuel use and the related greenhouse gas emissions and by setting limits on power consumption in buildings. It is not possible for one chapter to cover all of the legislation in a country, much less worldwide, but the following examples illustrate the range of legislation. Lighting practitioners need to be aware of the legislation in the locations where they practice to ensure their designs are compliant and don't delay the design, construction, or occupancy of the buildings they light.

In 2018 New York City passed Local Law 33 which requires most commercial buildings in the city to benchmark their building using ENERGY STAR Portfolio Manager (see Chapter 11) and to post the resulting score in a conspicuous location near the public entrance to the building. Benchmark scores must be submitted every year (NYC Buildings Energy Grades). Building owners will be expected to improve their score at designated intervals (such as each decade) or face fines or other penalties. The goal of the legislation is to increase the public's awareness of building energy efficiency and the efficiency of the stock of buildings in the city.

In 2023 the United Kingdom passed Energy Act 2023 with an overall goal of "net zero by 2050 in a pragmatic, proportionate and realistic way." It includes a wide range of provisions including merging certain energy networks to increase efficiency, providing incentives for heat pumps and smart meters, and licensing of CO_2 storage and nuclear fusion (Landmark Energy Legislation Becomes Law in the UK, 2023).

In 2024 the U.S. Department of Energy issued a determination that adoption of ANSI/ASHRAE/IES Standard 90.1–2022: *Energy Standard for Buildings, Except Low-Rise Residential Buildings* (Standard 90.1–2022) would increase average energy savings by 9.8% over the 2019 edition. As a result, individual states must begin to update their codes to meet or exceed Standard 90.1-2022 (Determination Regarding...., 2024).

12.9 Utility Rebates and Government Incentives

Utilities have offered rebates and financial incentives for homes and businesses to improve their energy efficiency for decades. A common type of incentive program on the 1990s urged replacement of A-lamps with compact fluorescent lamps. In the 2010s a common rebate program offered building owners money for every incandescent or fluorescent luminaire that was replaced with an LED luminaire.

As with legislation, the range of rebates and incentives is too broad to fully cover, but as an example, the rebates and incentives offered on the government website of New York State (USA) include (Lighting and Controls Programs & Incentives, 2024):

- financial support for energy audits of commercial buildings;
- payment for up to 70% of the cost of an energy efficient relighting project for small businesses;

- discounts on qualified LED lighting products purchased through participating distributors (see Qualified Products Lists in Chapter 11);
- rebates on the cost of LED lighting and controls.

12.10 Additional Value Considerations

There are aspects of a lighting and control system that don't have obvious benefits or costs, or that are obvious but difficult to quantify. However, being difficult to quantify doesn't mean that they don't impact the value of the lighting design to the building or space, the user experience, etc. While not formally a part of the calculations discussed in this chapter, an attempt may be made to include them through a percentage increase or reduction in the initial costs and/or annual operating costs (Livingston, 2022).

12.10.1 Warranty

As of early 2024, five-year warranties on new LED luminaires have become commonplace. A low-cost luminaire with a shorter warranty may indicate that the manufacturer has low confidence in the durability of the luminaire or its components. Such a luminaire may require more maintenance or have a shorter useful life than those with longer warranties. In such a case, this may be reflected in calculations by increasing the expected frequency and cost of maintenance.

12.10.2 Lighting quality

As discussed in Chapter 1, quality lighting has benefits to the owner and the occupants, which vary by the type and purpose of the space. Some of the benefits vs costs include:

> Well-shielded luminaires may be more expensive than poorly shielded luminaires, but they can reduce direct glare for occupants of a space, making it more pleasant. Well-shielded luminaires may also reduce veiling reflections on display cases, potentially increasing sales, and on computer screens, potentially increasing employee productivity.
>
> Including decorative luminaires in a design is more expensive than not including them, but they support and enhance the aesthetics of a space, potentially increasing occupant satisfaction, foot traffic, sales, or some other key measure used by the owner.

> Tunable white luminaires may be more expensive than single CCT luminaires, but they can support circadian entrainment, worker health and well-being, and overall employee satisfaction thus potentially reducing absenteeism and employee turnover.

As with warranty, luminaires that provide higher quality of light, reduce glare, or enhance ambience may be rewarded with an off-set to their higher cost versus low quality luminaires.

12.10.3 Environmental costs

Certain luminaire components, such as PCBs, have been found to be harmful and have been banned. Others, such as the mercury in fluorescent tubes, must be recycled in some locations. Additional materials may face local regulation regarding use, recycling, and/or disposal. Some materials may appear on warning lists, such as the Declare Red List discussed in Chapter 11, that owners want to avoid. In all of these cases, luminaires containing these materials may face high end-of-life disposal costs, which might be reflected in the life cycle cost analysis. Luminaires with RoHS, Declare, and other labels identifying hazardous or potentially hazardous materials may be preferable to those without for some projects.

12.10.4 Green programs

As discussed in the introduction to Chapter 11, in many cases building owners can charge higher rents in buildings that achieve certification in one of the green building programs (i.e., LEED, Green Globes, etc.). In some cases, the stringent requirements for achieving points in these programs can increase the cost of the lighting and control systems but may be rapidly off-set by the higher rents. An off-set may be added to any of the calculations above to account for the increased value delivered by more expensive luminaires.

References

Determination Regarding Energy Efficiency Improvements in ANSI/ASHRAE/IES Standard 90.1-2022 (2024). Available at: https://www.regulations.gov/document/EERE-2023-BT-DET-0017-0001. [accessed March 19, 2024].

IES Economics Committee (2020). *ANSI/IES RP-31-20 Recommended Practice: Economic Analysis of Lighting*. New York: Illuminating Engineering Society.

Landmark Energy Legislation Becomes Law in the UK (2023). Available at: https://knowledge.energyinst.org/new-energy-world/article?id=138372. [accessed March 19, 2024].

Lighting and Controls Programs & Incentives (2024). Available at: https://www.nyserda.ny.gov/PutEnergyToWork/Energy-Program-and-Incentives/Lighting-and-Controls-Programs-and-Incentives. [accessed March 19, 2024].

Livingston, Jason (2022). *Designing with Light: The Art, Science, and Practice of Architectural Lighting, 2nd Ed.* Hoboken: John Wiley & Sons, Inc.

NYC Buildings Energy Grades (n.d.). Available at: https://www.nyc.gov/site/buildings/property-or-business-owner/energy-grades.page. [accessed March 19, 2024].

Thumann, A., Neihus, T., Younger, W. (2013). *Handbook of Energy Audits, 9th Ed.* Lilburn, GA: The Fairmont Press, Inc.

Understanding the Anatomy of a Business Electricity Bill (2023). Available at: https://diversegy.com/anatomy-of-business-electricity-bill/. [accessed March 19, 2024].

13

Lighting Maintenance and Environmental Safety

Abstract

After a lighting system is installed it must be maintained. This chapter discusses the importance of luminaire maintenance, the steps to take when cleaning a luminaire and replacing the various lamp types, and possible causes of recurring lamp failure. It explains the maintenance strategies of spot relamping and group relamping. It ends with a discussion of the environmental concerns surrounding the disposal of lamps, ballasts, and luminaires.

Keywords: Luminaire cleaning, luminaire maintenance, lamp replacement, lamp failure, spot relamping, group relamping.

13.1 Introduction

When one considers that a short-lived lighting system will be in operation for five to ten years, and that many lighting systems are expected to operate as designed for 20 years or more, it becomes obvious that proper and planned maintenance of the lighting system is critical. If maintenance is not considered, several undesirable outcomes are possible, including.

> A lighting practitioner who knows that the lighting system will not be maintained may set a higher target light level to compensate for excessive recoverable light loss factors, as described in Chapter 8. This wastes energy and the owner's money.
>
> A lighting practitioner who disregards maintenance may specify luminaires that are difficult to maintain, install them in hard to reach or inaccessible locations, or specify overly expensive, difficult to obtain, or difficult to maintain components. All of these will result

in a lighting system that degrades over time unless the owner is willing to put in an inordinate amount of time, money, and effort to maintain the design.

A facilities manager who postpones or disregards lighting maintenance finds that over time the lighting system stops delivering the proper amount of light needed, where it is needed, when it is needed, and reliably.

Facilities managers and maintenance personnel who don't understand the design intent may keep luminaires operating, but over time inadvertently allow the light distribution or timing to change, altering the design and reducing its effectiveness and efficiency.

Occupants find the poorly maintained lighting distracting, uncomfortable, or unattractive, which can result in lower employee productivity, lower sales, loss of aesthetics, reduced safety and/or increased accidents or crime.

13.2 Cleaning and Relamping Basics

It is good practice to clean and inspect the luminaire every time a lamp is replaced (for replaceable lamp luminaires). This will ensure that the luminaire is operating as close to design intent as possible, and maintenance staff can identify small repair or replacement issues before they become worse. The typical cleaning and relamping steps are:

1. Turn off or disconnect power. For track heads, remove the head from the track if it does not have an integral switch. Energized lamp sockets are an electric shock hazard to personnel when cleaning and a spark hazard when replacing the lamp.
2. Survey the surrounding area. If the floor under the luminaire or the surrounding furnishings could be damaged by falling materials, which range from dust balls to dropped lamps to cleaning fluid, cover the area with drop cloths.
3. Carefully open the luminaire. Many luminaires have lenses, diffusers, baffles, or louvers (see Chapter 6) that must be removed to access the lamp. Be sure to understand how to remove these (twist and pull, release a latch, etc.) and do so carefully. Note the orientation of focusable luminaires so that the luminaire can be correctly refocused. Set removable components aside to be cleaned. Damaged components should be replaced.

4. Remove the failed lamp and set aside for disposal or recycling.
5. Clean the luminaire and lenses, etc. Some exterior luminaires can be cleaned with soapy water or a high-pressure hose. Interior luminaires should be wiped down, inside and out, with a soft cloth dampened with water or a mild cleaning solution. Keep in mind that detergents may cause streaking on glass or plastic reflectors, lenses, and diffusers, and that ammonia can cause permanent fogging of some plastics. Be sure to dry all components thoroughly. While the luminaire is open, inspect the socket(s) for damage. Note any repairs or replacements that need to be made and communicate them through proper channels.
6. Replace the lamp(s) and re-install the shielding components.
7. When re-energizing the lamp, observe its operation. Lamps that are delayed in restarting or that flicker may indicate the ballast or transformer is beginning to fail. Knowledgeable personnel will be able to determine if the restart is normal or a cause for concern.
8. If the luminaire has an emergency battery pack, press the test button to confirm that the battery is working. If it is not, communicate the need for a replacement through proper channels.

Note that luminaires with integral LEDs do not need lamp replacement, but it is good practice to clean them on a regular basis following the procedures outlined above.

13.3 Lamp Replacement by Lamp Type

Regardless of the lamp type, luminaires should be disconnected from power before lamp replacement. Failure to do so could result in electric shock to the maintenance worker, arcing between the socket and the lamp which could damage both, and/or tripping of the local circuit protection device.

13.3.1 Incandescent lamps

Failure of incandescent and halogen lamps does not strain related components. Replace incandescent lamps with a lamp of the same base, shape, wattage, and beam angle.

13.3.2 Halogen lamps

The quartz envelope of a halogen lamp should only be handled with a cloth, tissue, or gloves to avoid leaving fingerprint oils on its surface, which can result in premature failure of the new lamp. Replace the failed lamp with a lamp of the same base, shape, wattage, and beam angle.

13.3.3 Linear and compact fluorescent lamps

Fluorescent lamp failure can have several causes, and failure of one component of the system can stress other components causing them to fail prematurely. Failed fluorescent lamps should be handled appropriately. For example, some local governments require recycling of fluorescent lamps or treating them as hazardous waste because of the mercury content, which is discussed later in this chapter.

> Failed fluorescent lamps don't stop the ballast from operating. The ballast will continue to work at starting the lamp, reducing its life. Likewise, lamps that are flickering or glowing are stressing the ballast and reducing its life. Replace these lamps promptly with lamps of the same CCT and color rendering.
>
> In two lamp luminaires, a failure of one lamp will cause the second lamp to operate at reduced current, reducing its life. Replace the failed lamp promptly. If in doubt about the remaining life of the second lamp (if it shows heavy end blackening, for example), replace both lamps with lamps of the same CCT and color rendering.
>
> If replacing the lamps doesn't solve the problem, the ballast is the likely cause. Replace the ballast with the same type (instant start, rapid start, etc.). Check all sockets and wiring connections within the luminaire at the same time.

13.3.4 High-intensity discharge lamps

HID lamps can fail in several ways, depending on the lamp technology. High-pressure sodium lamps often cycle on/off at the end of their life. Metal halide lamps will usually stop starting. They may also show considerable color shift at the end of their life.

> Many HID lamps are sensitive to the mounting orientation. Use lamps designed for horizontal or vertical orientation in the appropriate luminaires, or use lamps designed for universal orientation.
>
> Replace HID lamps with a lamp of the same base, shape, wattage, CCT, color rendering, and (if applicable) beam angle.

If replacing the lamps doesn't solve the problem, the ballast is the likely cause. Replace the ballast with the same type. Check all sockets and wiring connections within the luminaire at the same time.

Note that there are caveats that apply only to HID lamps.

High-pressure sodium lamps require an ignitor in addition to the ballast. If the replacement lamp doesn't start, determine if the problem is the ignitor or the ballast.

Metal halide lamps may have noticeable color variation at the beginning of their life, which is unrelated to the color shifts that may occur at the end of their life. If a metal halide lamp produces an unexpected color when energized, allow a few days for the color to stabilize before assuming it is mislabeled or incorrectly manufactured.

13.3.5 LED lamps

Luminaires with integral LEDs do not need lamp replacement, and the LED module may not be replaceable without substantial disassembly of the luminaire. LED lamp issues are likely related to driver failure, and drivers are replaceable.

LED retrofit lamps are subject to failure, primarily because the small driver in the base of the lamp is susceptible to voltage variations and heat. A failed retrofit lamp cannot be repaired. Replace failed retrofit lamps with lamps of the same base, shape, wattage, CCT, and (if applicable) beam angle.

13.4 Relamping Strategies

Where luminaires with replaceable lamps are used, there are two relamping options – spot relamping and group relamping. Each has advantages and disadvantages that the lighting practitioner and maintenance personnel must consider to determine the most appropriate method for each space, floor, or building.

13.4.1 Spot relamping

Spot relamping is simply replacing lamps when they fail. With fluorescent lamps, spot relamping preserves the life of the ballast and other lamps in the luminaire. Spot relamping is the most effective way of maintaining the lighting design and illumination levels, but there can be drawbacks.

The first drawback is cost. The labor required to replace individual failed lamps can be considerable. One estimate is that it takes 10 hours to group replace 100 fluorescent lamps, but it takes 50 hours to spot replace the same number (IES Maintenance Committee, 2020).

The second potential drawback is mismatched replacement, especially with fluorescent lamps. Each fluorescent lamp manufacturer uses a different blend of phosphors to create light. The result is that even if two lamps have the same nominal CCT and color rendering, they can still have different chromaticities. In a room illuminated with lamps from manufacturer A, but with a few lamps replaced with manufacturer B, the replacement lamps could be quite noticeable resulting in a patchwork of colors instead of consistent white light.

Spot relamping should include the inspection and cleaning steps outlined earlier in this chapter.

13.4.2 Group relamping

Group relamping is replacing all of the lamps in an area at the same time at regularly scheduled intervals. Although it can be costly to purchase a replacement for every lamp on a floor or in a building, as noted earlier the cost savings on labor can be considerable and can off-set the cost of the lamps. It also has the advantage of minimizing disruption to the workers under the luminaires. One rule-of-thumb to determine if group relamping is the right choice is if the cost of labor to spot-replace one lamp, less the cost of group relamping it, is greater than the cost of the new lamp.

There are two common calculations that may be used to determine the group relamping interval. These may be modified based on the conditions and use of the space, budget, etc.

- When 5% to 10% of the lamps in an area have failed.
- At 70% of rated lamp life. As noted in Chapter 2, the life of fluorescent lamps varies from 15,000 to 40,000 hours, so it is important to know the projected life of the installed lamp, not a typical life for the lamp type.

Group relamping should include the inspection and cleaning steps outlined earlier in this chapter. The formula to calculate the group relamping interval eqn. (13.1) is:

$$\text{Group relamping interval} = \frac{(\text{average lamp rated life in hours}) \times (\text{rated life \% at replacement})}{\text{annual luminare operation in hours}}. \tag{13.1}$$

13.4.3 LED relamping and replacement

As noted earlier, LED modules rarely fail, so relamping is typically not considered. LED modules are most likely to fail at the beginning of their life and within the manufacturer's and contractor's warranty period, so repair or replacement should not be a burden on the owner. Failure of other components, notably the driver, may occur after several years, but drivers are intended to be replaceable, typically through the aperture after the trim is removed.

The larger replacement issue arises when the LEDs, through the expected process of lamp lumen depreciation (see Chapter 8) are no longer producing sufficient illumination and maintenance options are unavailable.

For example, a group of LED luminaires with an L70 of 50,000 hours that are in use for 14 hours per day would reach the calculated end of their useful life after 9.8 years as the light output slowly declines to 70% of initial brightness. Depending on the occupants and the application, complaints about the light level may arise sometime between 8 and 12 years after the installation. The owner may decide to replace the LEDs only to find that the original LED module has been out of production for many years. Contemporary replacement modules may not be available or suitable. At that point, the owner may have little recourse but to replace the luminaires.

This scenario illustrates why owners may require, and lighting practitioners may decide to specify, luminaires designed according to a standard that facilitates replacements, such as Zhaga. See Chapter 11 for more on this as an issue of sustainability.

13.5 Troubleshooting

When a luminaire isn't functioning properly maintenance personnel need to diagnose the problem in order to solve it. The general approach is to begin at the lamp (especially with traditional light sources) and work backwards. For example, if a luminaire with a traditional light source isn't working:

1. Replace the lamp. If this doesn't solve the problem:
2. Replace the luminaire's transformer or ballast. If this doesn't solve the problem:
3. Inspect socket(s) and internal wiring. Replace or repair as needed. If this doesn't solve the problem:
4. Are other luminaires in the zone working? If not, check circuit breaker, dimmer, 0-10 V output device, control system programming, and other control system related components.

13.5.1 Recurring failures

Recurring failures are likely due to outside factors that may take more time to isolate and remedy. Outside factors include:

Ambient temperature

- Fluorescent lamps are sensitive to low temperatures and may require a "cold-weather" ballast to start.
- At very low temperatures the output of fluorescent lamps may be reduced.
- High temperature can cause premature failure of transformers and ballasts.
- LEDs are sensitive to high temperature, so replacement with a different technology might be warranted.

Luminaire cooling

- Improper or blocked ventilation can increase the temperature inside a luminaire, causing problems that mimic high ambient temperature.
- Improper ventilation can cause a luminaire's internal thermal protector (if it has one) to turn the light off until the luminaire cools. This can result in on/off cycling.

Voltage regulation

- Consistent low voltage, such as during a brownout, can result in consistent low output, failure of the arc in fluorescent and HID lamps causing them to go out, or lamps not starting.
- Drops in voltage, such as when a nearby high-wattage device is turned on, can result in momentary drops on voltage producing the same results as consistent low voltage.
- Consistent over voltage can shorten the life of incandescent lamps.

13.5.2 LED retrofit lamps

LED retrofit lamps can be particularly sensitive to changes to the sine wave produced by a dimmer. This can often result in flickering, strobing, and/or turning off when dimmed. Manufacturers may change components, especially the driver or driver components, without changing product names or catalog numbers, giving the impression that consumers are purchasing identical lamps over time. If a new retrofit lamp isn't performing as expected a different lamp may solve the problem. Many retrofit lamp manufacturers list dimmers that have been tested and found to be compatible with a given lamp. Note, however, that "compatible" doesn't necessarily mean "will dim smoothly to the desired low output." Several lamps may need to be tested before a suitable replacement is found.

13.6 Environmental Concerns

Luminaires and their components can contain materials that can be hazardous to human health. How they are dealt with during repair or disposal depends greatly on the material and the rules and regulations of the locale. Below are some examples.

13.6.1 Luminaire components

Luminaires are complex assemblies that may contain metal, glass, plastics, and electronics, each of which may be made of several components. Ingredient health disclosure labels, such as Declare, list all product ingredients. See Chapter 11 for additional information.

13.6.2 Mercury-containing lamps

As explained in Chapter 4, fluorescent lamps require mercury to produce the ultraviolet light that phosphors convert into visible light. Mercury is also present in many HID lamps. Short-term exposure to mercury may cause irritation to the eyes, skin, and stomach, cough, chest pain, or difficulty breathing (Mercury, 2019). Long-term exposure can cause tremors, neuromuscular effects, and cognitive and motor disfunction (Mercury and Health, 2017). While fluorescent lamps contain only trace amounts – a CFL lamp contains 1/100th the mercury in a thermometer (Cleaning Up a Broken CFL, 2023) – they still must be treated with care.

In the United States, mercury-containing lamps are classified and treated according to their mercury content. Lamps that pass the U.S. Environmental Protection Agency (EPA) Toxicity Characteristic Leaching Procedure (TCLP) are very low in mercury and may be disposed of in landfills, local regulations permitting. Generally, lamps containing less than 4 mg of mercury will pass the TCLP test. These lamps often have green end caps.

Lamps with higher levels of mercury may be treated as "universal waste" where the waste disposal company has specific requirements on how to handle, ship, and dispose of mercury-containing waste. Lamps with the highest levels of mercury may need to be treated as hazardous waste. Many waste haulers provide special holding containers for customers who produce mercury-containing waste.

The mercury in lamps, especially fluorescent lamps, can be extracted and recycled. Lamp recycling is required in some locations, and manufacturers generally support lamp recycling and mercury recovery efforts. In the United States, some recyclers provide specially-lined cardboard containers to ship lamps to a recycling facility.

Although a broken fluorescent lamp releases only a small amount of mercury, mercury is toxic enough for it to be a concern. The general procedure for cleaning up a broken lamp is (Cleaning Up a Broken CFL, 2023):

1. Have people and pets leave the room.
2. Shut off any forced air heating/air conditioning and open a window to the outdoors, if possible, for 5–10 minutes to air out the room.
3. Do not vacuum. Wear gloves to pick up the broken glass and phosphor powder. Use a wet cloth or tape to gather as much as possible.
4. For residential waste, place the debris in a sealed container and recycle or dispose of with other waste, according to local requirements.
5. For commercial waste, collect the debris in a clearly marked container for hazardous waste disposal, if required. Generally, broken lamps can be stored together in one container for up to one year.
6. Dispose of gloves and thoroughly wash hands.
7. Air out the room for several hours or as long as practical.

13.6.3 PCB-containing ballasts

Polychlorinated biphenyls (PCBs) were used in ballasts for many years because of their electrical insulating properties. Unfortunately, they were also discovered to be carcinogens that also have negative reproductive, neurological, and endocrine effects. In the United States they were banned in 1979 (Learn About Polychlorinated Biphenyls, 2023).

Ballasts have very long lives, although any PCB-containing ballasts remaining in buildings are past their designed life span. However, some buildings may still have PCB-containing ballasts. PCB-containing ballasts can leak small amounts of PCBs into the air during normal use. Therefore, the EPA recommends that buildings replace PCB-containing ballasts as soon as possible. According to the EPA Toxic Controlled Substances Act, PCB-containing ballasts must be recycled or incinerated (Disposal of Fluorescent Light Ballasts FLB, 2023).

13.7 Class 1 and Class 2 Wiring

Commercial buildings typically contain two categories of wire installations, which are categorized by *NFPA 70: National Electrical Code* as Class 1 and Class 2. Class 1 is for wiring that carries up to 600 V and is commonly referred to as line voltage wiring – that is, wiring that carries the local nominal voltage (120 V in North America and 230 V in most of the rest of the world). The voltage in Class 1 wiring is high enough to be a fire and shock hazard. Building and electrical codes set appropriate requirements for the wiring composition, wiring techniques, circuit protection, etc. Installers of Class 1 wiring must be properly trained and, depending on the locale, may need to be licensed.

Class 2 wiring carries up to 30 V and typically describes data and control wiring. Class 2 wiring is not usually be considered a fire or shock hazard. While installers of Class 2 wiring should be properly trained, in many locales they are not required to be licensed.

References

Cleaning Up a Broken CFL (2023). Available at: https://www.epa.gov/mercury/cleaning-broken-cfl. [accessed on March 26, 2024].

Disposal of Fluorescent Light Ballasts (FLB) (2023). Available at: https://www.epa.gov/pcbs/disposal-fluorescent-light-ballasts-flb. [accessed March 26, 2024].

Early, Mark, Ed. (2016). *National Electrical Code Handbook, 14th Ed.* Quincey, MA: National Fire Prevention Association.

IES Maintenance Committee (2020). *ANSI/IES RP-36-20 Recommended Practice: Lighting Maintenanc*e. New York: Illuminating Engineering Society.

Learn About Polychlorinated Biphenyls (2023). Available at: https://www.epa.gov/pcbs/learn-about-polychlorinated-biphenyls. [accessed March 26, 2024].

Livingston, Jason (2022). *Designing with Light: The Art, Science, and Practice of Architectural Lighting, 2nd Ed.* Hoboken: John Wiley & Sons, Inc.

Mercury (2019). Available at: https://www.cdc.gov/niosh/topics/mercury/default.html#:~:text=Some%20of%20the%20health%20effects,harmed%20from%20exposure%20to%20mercury. [accessed March 06, 2024].

Mercury and Health (2017). Available at: https://www.who.int/news-room/fact-sheets/detail/mercury-and-health. [accessed March 26, 2024].

14

Lighting Audits

Abstract

Before undertaking a lighting retrofit or replacement, the existing lighting system must be audited to understand what is installed. This chapter begins with an overview of the reasons an owner may undertake a lighting audit. It explains the four levels of audits and their increasing complexity, and provides details on lighting audit procedures and note taking at each level. The chapter ends with an outline of a typical audit report.

Keywords: Audit, Level 1 audit, Level 2 audit, Level 3 audit, retrofit, retrofit, redesign, benchmark, ENERGY STAR.

14.1 Introduction

Replacing the lighting systems of older buildings that are illuminated with fluorescent, HID, or even incandescent lighting offers the opportunity for energy savings, reduced environmental impact, and improved lighting quality. In order for the owner to understand the impact of a lighting renovation the existing lighting system needs to be surveyed so it can be analysed and compared to one or more proposed systems.

The two broadly used standards for audits are *ISO 50002:2014 Energy Audits – Requirements with Guidance for Use* and *ANSI/ASHRAE/ACCA Standard 211-2019 (RA2023) Standard for Commercial Building Energy Audits*. There are similarities and overlaps between them. This chapter discusses general audit procedures and data collection, not one standard or the other. Financial analysis is covered in Chapter 12.

14.2 Lighting Audit Reasons and Results

There are many reasons a building owner will undertake an audit of the lighting system. The desire to reduce utility costs is a common reason. Other reasons include: occupants may be complaining of spaces that are too dark, too bright, or that have excessive glare; excessive maintenance costs for outdated technology; luminaires may visibly be reaching the end of their useful life.

Once the decision is made to undertake an audit, the goal is often to determine if the lighting improvement strategy is one of retrofitting, replacement, or redesign.

> Retrofitting is installing new components into an existing luminaire. It could be simply replacing outdated fluorescent lamps with retrofit LED lamps. A more extensive retrofit might also replace the ballast and the lens to create something close to a new luminaire. If the goal is simply to reduce utility costs a retrofit makes sense.
>
> Luminaire replacement typically means a one-for-one replacement of existing luminaires with new luminaires of similar or better performance. Luminaire replacement is indicated if the installed luminaires are clearly past the end of their life. Replacement is usually less expensive than redesign, but won't capture all of a redesign's potential benefits.
>
> Redesign starts with a blank slate to develop a new lighting design that takes advantage of current luminaires and technology to meet the needs of the occupants and their visual tasks, and to maximize energy efficiency. It is the most expensive option, but may also provide the greatest returns on savings, lighting quality, and overall occupant satisfaction.

14.3 Audit Types or Levels

Audits can be described as Type or Level 0, 1, 2, or 3. Each successive number providing a more thorough, detailed analysis. The Types/Levels are nested so that higher levels include the work of lower levels (Thumann, et al.). In this chapter audits will be described with the word Level.

14.3.1 Level 0 audit

A Level 0 audit benchmarks the building or the building system being audited. As discussed in Chapter 11, in North America the ENERGY STAR Portfolio Manager is a free, online benchmarking tool of the US EPA. Benchmarking measures the current performance of the building against similar buildings.

Generally, benchmarking requires energy use data for at least 12 full consecutive months. This gives the owner an understanding of where their building is in the spectrum of existing buildings, and the energy savings potential of a renovation or retrofit.

14.3.2 Level 1 audit

A Level 1 audit is also referred to as a walk-through audit. It begins with an analysis of the energy bills and includes a tour of the building to visually survey the existing lighting system. The Level 1 audit will identify the most easily remedied inefficiencies, such as inoperable occupancy detectors and continuing use of inefficient incandescent lamps. It may also suggest a more thorough audit is needed.

14.3.3 Level 2 audit

A Level 2 audit includes an energy survey and analysis, with a more detailed review of luminaires and the control system's programming and function. It may also include measurements of intensity for comparison to recommended light levels, measurements of color rendering, and other measurements the auditor finds relevant to the owner's goals.

14.3.4 Level 3 audit

A Level 3 audit is the most intensive, and may include a computer simulation of the energy use of the building to test various renovation or replacement strategies and thorough budgeting of the renovation or replacement.

14.4 The Lighting Audit Process

Audits that focus only on the lighting system are relatively easy compared to full building energy use audits. Such an audit is generally undertaken with the goal of determining the extent of an energy-efficient lighting retrofit or renovation. Is the installed lighting system a candidate for replacing the existing lamps with LED retrofits? Are there spaces where the luminaires should be replaced on a one-for one basis? Do the potential energy savings, and/or improved lighting quality (and/or aesthetics) merit a completely new lighting system?

The first step in a lighting audit is determining who is on the audit team (if it is performed internally) or hiring of a professional auditor. The second step is to notify those who will be asked to assist in the audit. For example, someone from the building security or maintenance staff may need to accompany the auditor(s) to unlock doors. A building engineer or electrician may need to accompany the auditor(s) with a ladder to access luminaires and report on the lamps in use.

The third step is to notify the building occupants that people will be entering every space, examining the lighting system and controls, taking measurements, perhaps turning lights on and off to understand the switch wiring, and possibly asking questions about the use of spaces or the lighting system. Building occupants should be assured that intrusion and disruptions will be kept to a minimum.

The fourth step is to perform the audit, depending on the level, as described below.

14.4.1 Level 0 lighting audit

Lighting systems are rarely metered separately from the rest of a building's electrical system. Therefore, conventional benchmarking of a lighting system is usually not possible. However, energy use information can be captured by combining a Level 1 audit with occupancy/light loggers (Figure 14.1). These battery-powered devices are temporarily mounted to a wall in a room. They sense occupancy and light, and record both for later download and analysis. Installing loggers in a set of typical rooms in a building, combined with a knowledge of the lighting load in the rooms, can be the basis for an estimate of the lighting system's power consumption.

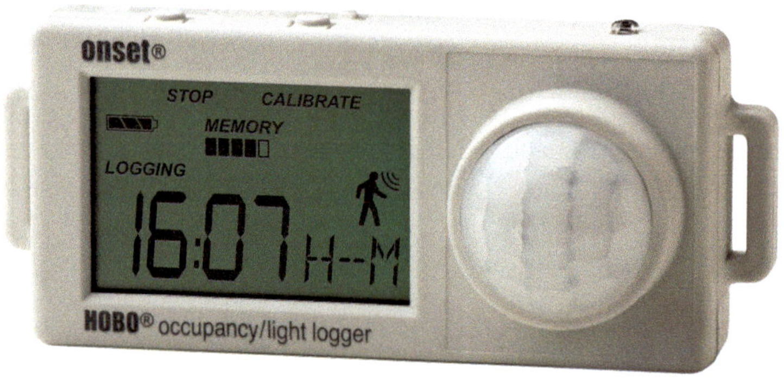

Figure 14.1 Occupancy/lighting logger.
Courtesy of Onset HOBO Data Loggers

In addition, the owner of multiple buildings of a similar type might perform Level 1 audits on multiple buildings in their portfolio to generate an internal benchmark and the range between the best and worst performing buildings.

14.4.2 Level 1 lighting audit

The Level 1 audit should begin with a clear set of procedures for conducting the audit, an understanding of the data that should be collected, and the format for data collection. This information should be conveyed to the relevant parties before the audit begins. The auditors can be greatly assisted by copies of the original drawings, schedules, and luminaire cut sheets. Lighting or electrical drawings are the most useful, but even floorplans that can be marked up during the audit are helpful.

The Level 1 audit is a walk-through, not an in-depth analysis. The goal is to gather general information on the lighting system. The information gathered is usually the number of luminaires, the number and type of lamps, the lamp wattage, controls, and the room dimensions (ASHRAE Standing Standard Project Committee 211, 2023). The walk-through audit should count and describe all ceiling-mounted luminaires, wall sconces and other permanently wired decorative luminaires, exit signs, and emergency lights.

Recording of information could take the form of notes and narratives, or checklists. A note/narrative format may be one or two simple sentences about each space. For example:

> Corridor 101. 6'W × 80'L. 12 @ 6" diam. downlights, 13W CFL lamp. No switches. Eng. reports lights on timer – on 6 am, off midnight.
>
> Office 105. 14'W × 16' L. 4 @ 2 × 2 fluorescent, 2 × 32W U-lamps. 1 switch. No sensors.

A checklist would gather similar information (Table 14.1).

Table 14.1 Level 1 audit form.

Room	Dimensions	Operating hours	Luminaire type and quantity	Lamps/ wattage	Measured light level	Controls
101	6′ × 80′ × 10′	6 am–midnight	12 6″ dia. downlights	13 W CFL	10 fc	Timer
105	14′ × 16′ × 10′	9 am–5 pm	4 2 × 2 downlights	(2) 32 W u-bend	35 fc	On/off switch

Either approach would provide sufficient information to later compile an estimate of the load and controls in each room. These could be compared to the current energy conservation code, for example, to assess potential savings from a one-for-one system replacement or a full redesign.

14.4.3 Level 2 lighting audit

A Level 2 audit collects the same information as a Level 1 audit but supplements it with measurements of illuminance at task surfaces, CCT, CRI and/or TM-30, and other relevant measurements. If retrofit lamps are under consideration sample luminaires should be opened to get information about the ballasts that are installed. Other, less common, measurements may include photosynthetic photon flux density (PPFD) a measure of the amount of light reaching a surface based on the spectrum needed for plant photosynthesis, which would be relevant in spaces with many plants or a green wall. Note that most meters, regardless of what they are designed to measure, will need regular recalibration to ensure their measurements are accurate.

The control system is also inspected in a Level 2 audit. Occupancy sensors are tested to see if they are functioning, and a sample of them may be checked for their timeout settings. Pushbutton controls are tested to confirm they are communicating with the central controller. Presets are observed to confirm they align with their description and intended use. Table 14.2 is an example of a Level 2 survey form.

The information gathered in a Level 2 audit can be synthesized to determine the installed connected load and energy consumption, compare the installed load and control system functions to the current energy conservation code requirements, and compare existing light levels to current illuminance recommendations. The calculations found in Chapter 12 can then help determine if the lighting system should be replaced on a one-for-one basis or with a new design, the amount of energy that could be saved, and the resulting time period needed for the energy savings to pay for the lighting renovation.

14.4.4 Level 3 lighting audit

As noted earlier, a Level 3 lighting audit is rarely undertaken. The Level 2 audit energy use analysis combined with an energy savings or life cycle cost analysis using estimated luminaire quantities and costs is usually sufficient to prove that a retrofit or redesign should (or shouldn't) go forward. In this scenario, the equivalent of a Level 3 audit would be to fully document the retrofit or redesign so that exact manufactures, model numbers, and quantities

Table 14.2 Level 2 audit form.

Room	Dimensions	Operating hours	Luminaire type and quantity	Lamps/wattage	Total load	Ballast/transformer	Light level	CCT/CRI/TM-30	Notes

could be used to develop a more detailed and accurate budget and life cycle cost analysis.

14.5 Reports

The ISO and ANSI standards both include recommendations for the outline and content of a final energy audit report. One possible outline for a lighting only version is below.

I. Executive summary. A one-to-three page summary that includes the extent of the audit, the procedures, the analysis, the results, and any recommendations. Recommendations may be stated in annual energy savings, installation costs and return on investment, improvement in ambience, potential increase in sales or employee performance, etc.

II. Introduction
   - A. Description of the facility and the reasons for the audit
   - B. Audit scope, plan, and processes

III. Audit data and findings
   - A. Luminaires in use, including type, quantity, load, efficacy, etc.
   - B. Current energy consumption
   - C. Compliance issues, such as failed exit signs or inadequate illumination along egress paths

IV. Recommendations
   - A. Energy savings opportunities
   - B. Recommended energy savings strategy and calculations
   - C. Recommendations for code compliance

V. Appendices (optional)
   - A. Audit plan and collected data
   - B. Prints of energy bills, analysis and calculations, etc.
   - C. Drawings and/ or cut sheets of example or recommended retrofit kits, lighting plans, control system components, etc.

References

ASHRAE Standing Standard Project Committee 211 (2023). *ANSI/ASHRAE/ACCA Standard 211-2019 (RA2023) Standard for Commercial Building Energy Audits*. Peachtree Corners, GA: ASHRAE

Technical Committee ISO/TC 242 Energy Management (2014). *ISO 50002:2014 Energy Audits – Requirements with Guidance for Use.* Geneva: International Organization for Standards.

Thumann, A., Neihus, T., Younger, W. (2013). *Handbook of Energy Audits, 9th Ed.* Lilburn, GA: The Fairmont Press, Inc.

Glossary

Accent Lighting

Directional lighting to emphasize a particular object or surface feature, or to draw attention to a part of the field of view.

Adaptation

The process by which the eye becomes accustomed to more or less light than exposed to in the preceding period, resulting in changes to the sensitivity to light.

Aiming Angle

The angle, measured in degrees, between vertical and the centerline of the beam of light illuminating an object.

Ampere

The unit of measurement for electric current; it is the amount of electricity moving past a given point in a circuit during a specific period, also referred to as amp. Amperage is equal to wattage divided by voltage.

Angle of Incidence

The angle between a ray of light striking a surface and a line perpendicular to that surface.

Annual Operating Costs

The costs of electricity, maintenance, and other expenses related to a lighting system.

ANSI

Abbreviation for American National Standards Institute

Aperture

The opening in a luminaire through which light exits.

Arc Tube

A tube made of glass, clear quartz, or ceramic that contains the arc stream in an HID lamp.

ASHRAE

Abbreviation for American Society of Heating, Refrigerating and Air-Conditioning Engineers.

Astronomical Time Clock

A component of a lighting control system that is set to the local time, day of year, and geographic location, allowing it to trigger lighting changes based on time of day or based on astronomical events such as sunset or sunrise.

Baffle

A vertical blade at the aperture of a luminaire used to shield the source from direct view, to block light distribution at high angles to prevent glare, or to reflect and redirect light.

Ballast

A device used to operate fluorescent and HID lamps; the ballast provides the necessary starting voltage and regulates the current during operation.

Ballast Factor (BF)

The percentage of the rated lamp lumens that will be produced when a fluorescent lamp is operated on a given ballast compared to when operated on a standard or reference ballast.

Base

The part of a replaceable lamp that fits into a luminaire's socket.

Beam Angle

The angle through the center of the beam where the light level has fallen to 50 percent of the beam's maximum brightness (typically at the center of the beam). The beam angle may also be given for two planes at 90 degrees to one another when the beam is other than a simple cone.

Binning

The sorting of LEDs by characteristics such as of color, photometrics, and electrical performance.

Blackbody Locus

See Plankian Locus

Blackbody Radiator

A theoretical object that is a perfect absorber of all energy that strikes its surface, and that is an ideal emitter of energy whose spectral power distribution curve is based on its temperature. Also called Planckian radiator.

Candela (cd)

Unit of luminous intensity, describing the intensity of a light source in a specific direction equal to one lumen per steradian.

Candlepower (cp)

An obsolete measure of luminous intensity of a light source in a specific direction. It has been replaced by the candela. 1 cp = .98 cd

Candlepower Distribution Curve

A curve, plotted on polar coordinates, illustrating the distribution of luminous intensity of a lamp or luminaire in a plane through the light center

Chroma

The saturation or purity of a color

Chromatic Adaptation

An observer's ability to perceive colors of objects independent of changes in the light source.

Chromaticity

Refers to the color of a light source independent of its brightness

Circadian

Variations with a cycle of approximately 24 hours

Clerestory

The upper portion of a wall containing openings to admit daylight into a building

Coefficient of Utilization (CU)

The fraction of lamp lumens that reach the work plane directly from the light source and from inter-reflections

Color

The characteristic of visual perception that can be described by attributes of hue, brightness (or lightness), and colorfulness (or saturation or chroma). See Light Source Color and Object Color.

Color Fidelity

The degree to which an object's color looks similar to its appearance under a reference illuminant.

Color Rendering

The effect of an illuminant on the color appearance of objects by conscious or subconscious comparison with their color appearance under a reference illuminant.

Color Rendering Index (CRI)

A measure of the degree of color shift that objects undergo when illuminated by a light source, as compared with the color of those same objects when illuminated by a reference source of comparable color temperature, according to *CIE 13.3 Method of Measuring and Specifying Colour Rendering Properties of Light Sources.*

Color Temperature

A numerical indication of the warmth or coolness of white light; specifically, it is the absolute temperature of a blackbody radiator having a chromaticity equal to that of the light source, expressed in Kelvin. Strictly speaking, this applies only to incandescent light sources. Also see Correlated Color Temperature.

Colorimetry

The science of the measurement of color.

Commissioning

The process of setting, refining, and/or resetting a building system's operation; in lighting, this involves setting occupancy, vacancy, and photo sensors and control system presets.

Compact Fluorescent Lamp (CFL)

A small diameter fluorescent lamp that is folded, bent or spiralled to reduce length.

Cones

Three distinct retinal photoreceptors that provide color vision in response to high illuminance levels and that perceive color and detail. So called because their shape resembles a cone.

Contrast

The relationship between the luminance or color of an object and adjacent objects or its surround

Control Station

A wall mounted push-button, fader, or touchscreen use to control the lighting system.

Control Zone

A group of luminaires performing the same task and controlled in unison. Also called a lighting control zone or zone.

Cornea

The clear covering over the iris and pupil; the cornea helps to focus light onto the retina.

Correlated Color Temperature (CCT)

For non-incandescent sources, the temperature of a blackbody whose chromaticity most closely matches that of the light source.

Cove Lighting

A lighting technique in which a luminaire is set on a shelf or in a recess below the ceiling; light from the fixture is directed out of the recess to reflect off of the ceiling and into the illuminated space.

Current

The flow of electricity past a given point. Current is measured in amperes or amps.

Cut-Off Angle

The angle from a fixture's vertical axis at which a reflector, louver, or other shielding device cuts off direct visibility of the light source.

Daylight

Light from the sun and sky.

Daylighting

The intentional use of daylight as a significant daytime light source for a building's interior.

Daylight Harvesting

An energy-saving technique used in areas with significant daylight contribution in which a dimming system responds to a photocell by reducing the output of the luminaires when daylight is present; as daylight levels increase, electric light intensity decreases.

Daylight Zone

The area within a building that is adjacent to or beneath windows, clerestories, roof monitors, and/or skylights as defined by the applicable energy conservation code.

Delivery Charges

An item on a utility bill representing the costs for the utility to send electricity to a home or business.

Demand Response

Reduction of electricity usage during peak periods in response to a power company's request.

Diffuse

Term describing dispersed light distribution; it refers to the scattering or softening of light.

Diffuser

A translucent piece of material that shields the light source in a luminaire; to reduce glare or soften shadows. The light transmitted through the diffuser is redirected and scattered. The diffuser is usually perceived as less bright than the light source because the light is distributed over the larger area of the diffuser.

Digital Addressable Lighting Interface (DALI)

A communications protocol and network in which each luminaire has a distinct address, receives dimming information, and can report back to the lighting controller

Dim-to-Warm Lighting

Describes luminaires that automatically lower CCT as they are dimmed to emulate incandescent lighting.

Dimmer

An electrical device that controls luminaire intensity by regulating the flow of electricity to the light source.

Direct Distribution

Describes light leaving a fixture and entering the illuminated space without first bouncing off of any architectural elements such as the ceiling or walls

Direct Glare

Glare produced by a direct view of light sources; it is often the result of fixtures with inadequate shielding.

Direct/Indirect

A light distribution pattern or luminaire in which 40–60 percent of the light is directed downward and 40–60 percent of the light is directed upward

Discomfort Glare

Glare that produces discomfort but does not necessarily interfere with visual performance or visibility.

Distribution

1. The way light is emitted from a luminaire (upward, downward, etc)
2. The way light is spread over an area or throughout a space

Downlight

A type of ceiling luminaire, usually recessed, from which the light is directed downward

Driver

The electronic device regulating electricity to an LED; it is also known as a power supply.

Duv

A measure of the distance from the blackbody locus in the CIE 1960 (u, v) UCS diagram. A positive value is above the blackbody locus and a negative value is below the blackbody locus.

Efficacy

The effectiveness of a device that has different forms of energy input and output; for example, how well a light source receives an input of electricity and converts it to light output is measured as its efficacy.

Efficiency

The effectiveness of a device that has the same form of energy input and output. For example, how well transformers and ballasts receive an input of electricity, modify the electricity and output electricity is measured as their efficiency.

Electromagnetic interference.

High frequency interference (electrical noise) caused by electronic components or lamps that interferes with the operation of electrical equipment.

Electromagnetic Spectrum

The continuous range of electric and magnetic radiation encompassing all wavelengths (or frequencies).

Electronic Ballast

A ballast that uses semi-conductor components to increase the frequency of electricity for fluorescent and HID lamp operation (typically in the 20–40 kHz range); smaller inductive components provide the lamp current control. System efficiency is increased due to high frequency lamp operation.

Environmental Product Declaration

A third-party verified document that conveys the content and source of materials of a product.

Field Angle

The angle through the center of the beam where the light level has fallen to 10 percent of the beam's maximum brightness (typically at the center of the beam)

Fixture

See Luminaire

Flange

A horizontal ring at the bottom of a reflector that hides the gap between the ceiling material and the reflector. Also called the Overlap Flange.

Flicker

Fluctuation or modulation of luminance or spectrum over time.

Fluorescent Lamp

A light source consisting of a glass tube filled with argon, krypton, or other inert gas along with mercury; when electrical current is applied, the arc through the tube causes the mercury to emit ultraviolet radiation that excites the phosphors inside the lamp wall, causing them to radiate visible light.

Footcandle (fc)

The US Customary unit of illuminance; one foot candle is equal to one lumen per square foot.

Glare

The effect of brightness or differences in brightness within the visual field sufficiently high to cause annoyance, discomfort, or loss of visual performance.

Group Relamping

Replacing all of the lamps in an area at regularly scheduled intervals.

Halogen Lamp

An incandescent lamp with a lamp envelope made of quartz and a fill gas of one or more halogens (namely iodine, bromine chlorine, and fluorine), which slow the evaporation of the tungsten; It is also commonly called a quartz lamp or tungsten halogen lamp.

Heat Sink

A component of LED luminaires that draws head away from the LED and dissipates it.

Hertz (Hx)

Cycles per second

High Intensity Discharge (HID)

An electric discharge lamp in which an electric arc through a gas-filled chamber produces light without intermediate steps; also, it is a generic term describing mercury vapor, metal halide, high pressure sodium, and low-pressure sodium light sources and luminaires.

High Pressure Sodium Lamp (HPS)

A high intensity discharge (HID) lamp whose light is produced by radiation from sodium vapor

Housing

The outer enclosure of a luminaire.

Hue

The general color attribute of red, blue, yellow, green, etc.

Hunt Effect

A color appearance phenomenon where colorfulness of chromatic objects increases with luminance (even though chromaticity remains unchanged).

Illuminance

The light incident on a surface or plane; illuminance is commonly referred to as light level or brightness. It is expressed in footcandles or lux.

Index of Refraction

A measurement of the light bending properties of a material calculated as the speed of light in a vacuum divided by the speed of light in the material.

Indirect

A light distribution pattern or luminaire in which all of the light is directed upward to bounce off of the ceiling

Ingredient Health Disclosure

A form that lists the content of a product and may include the possible health affects of the contents.

Initial Cost

The cost to purchase and install a lighting system.

Initial Lumens

The light output of a light source when new.

Incandescent

Emitting light as a result of being heated.

A lamp technology that produces light by passing electricity through a tungsten filament, heating it to a high temperature and causing it to emit light.

Intensity

The amount of light energy emitted from a light source or delivered to a surface. Commonly referred to as brightness.

Interreflection

The multiple reflections of light by various room surfaces.

Intrinsically photosensitive retinal ganglion cell (ipRGC)

A light sensitive cell in the retina primarily responsible for circadian entrainment.

Kilowatt (kW)

1,000 watts

Kilowatt Hour (kWh)

A measure of electricity defined as a unit of work or energy, measured as 1 kilowatt of power expended for 1 hour.

Lamp

A manufactured light source.

Lamp Life

The average life of a lamp, in hours. For non-LED light sources lamp life is the point at which half of a group of test lamps will have failed and is typically based on an operation cycle of three hours on, then off.

For LEDs lamp life is based on a projection of the number of hours before light output has fallen a specified amount, usually 70 percent of initial brightness, which is described as L70. Testing is done according to standards set by the IES.

Layering

Using more than one lighting technique in a space to achieve multiple goals, create visual interest, and/or give users greater control.

LED Package

An assembly of one or more LEDs with an optical element and provisions for thermal, mechanical and electrical connections.

Life-Cycle Cost

The total costs associated with purchasing, operating, and maintaining a system over the life of that system.

Light

The portion of the electromagnetic spectrum, ranging from about 380 to 780 nanometers, that is capable of exciting the retina and producing visual sensation.

Light Emitting Diode (LED)

A solid-state light source that emits ultraviolet, visible or infrared light depending on its composition.

Light Loss Factors (LLFs)

Factors that account for a lighting system's operation at less than initial conditions; these factors are used to calculate maintained light levels. LLFs are divided into two categories, recoverable and nonrecoverable. Examples are lamp lumen depreciation and luminaire surface depreciation.

Lighting Power Density (LPD)

The watts per square foot or square meter of power used by, or allocated to, lighting; energy conservation codes allocate LPDs based on space type or building type.

Light Source

Any object that creates light. Often use to refer to the technology used to produce light or to the source of light in a luminaire.

Louver

See Baffle

Low Voltage Lamp

A lamp (typically compact halogen) that operates at lower than 120V, typically 12V, and requires the use of a transformer

Lumen (lm)

A unit of light flow, or luminous flux; the lumen rating of a lamp is a measure of the total light output of the lamp.

Luminaire

A complete lighting unit consisting of a light source, along with the parts designed to hold the light source, connect the light source to power, and distribute the light; also called a fixture.

Luminance

The brightness of a light source or of light reflected from an illuminated surface. It is expressed as candelas per square foot or square meter.

Luminous

Emitting light

Lux (lx)

The International System of Units (SI) unit of measure for illuminance of a surface; one lux is equal to one lumen per square meter. One lux equals 0.093 foot candles.

Maintained Illuminance

The calculated light levels of a space after factoring in light loss factors such as lamp lumen depreciation, luminaire dirt depreciation, and room surface dirt depreciation.

Matte

Describes a diffuse surface finish that reflects light uniformly in all directions.

Mean Lumens

The average light output over the life of a lamp

Melanopsin

The pigment found in the intrinsically photosensitive retinal ganglion cell (ipRGC)

Mesopic Vision

Vision utilizing both rods and cones; mesopic vision occurs at illuminance levels between .1 and 100 lx.

Metal Halide

A type of high intensity discharge lamp in which light is produced by passing an electric arc through a combination of metal halide and mercury vapors

Movement

A change in the intensity, color and/or distribution of light over time

Nadir

A reference direction directly below a luminaire, or "straight down" (0 degree angle).

Narrow Band Emitter

Another name for LEDs, used because they only emit a portion of the visible spectrum.

Occupancy Sensor

A device that automatically turns lights on when motion is detected and automatically turns lights off after a set time when motion is no longer detected; it may use ultrasonic, microphonic, and/or infrared technologies.

Opaque

Describes a material which blocks the passage of light

Optics

A term referring to the components of a light fixture (such as reflectors, refractors, lenses, louvers) or to the light emitting or light controlling performance of a fixture

Organic Light Emitting Diode (OLED)

A thin-film light emitting diode composed of carbon-based material layered between two electrode surfaces and which is a diffuse area light source.

Parabolic Aluminized Reflector (PAR)

A sealed lamp unit having a parabolic shaped reflector, light source (typically halogen or metal halide), and a lens; PAR lamps are available in a variety of beam angles.

Pendant

Describes a luminaire hanging from the ceiling

Photocell

A light-sensing device used to control luminaires and dimmers in response to detected light levels.

Photometric Report

A set of printed data describing the light distribution, efficiency, and zonal lumen output of a luminaire

Phosphor

A substance that emits visible light when excited by radiation. In fluorescent lamps the radiation is UV light. In most LEDs the radiation is blue light.

Photon

A particle representing a quantum of light or other electromagnetic energy. A photon carries energy but has zero mass.

Photopic Vision

Vision resulting from the cone cells of the retina; generalized as daytime vision, photopic vision occurs at illuminance levels greater than 100 lx.

Plankian Locus

The locus of points on a chromaticity diagram representing the chromaticities of a blackbody at various temperatures.

Power Over Ethernet (POE)

A technology that permits multi-conductor data wiring to also carry electricity to power devices defined in IEEE 802.3-2018 Standard for Ethernet.

Preset

A set of light levels saved into the memory of a lighting controller for recall when needed.

Pupil

The opening of the iris that admits light into the eye

Qualified Products List

A list of luminaires that have met the requirements for a particular purpose, such as energy efficiency.

Rate of Return

The percentage of an investment that is recovered each year.

Recessed

Describes a luminaire that is installed so that the housing of the luminaire is above the surface of the ceiling, inside a wall, or below the floor

Reflectance

The ratio of light reflected from a surface to the light incident on the surface; reflectances are often used for lighting calculations.

Reflected Glare

The reflection from a specular or semi-specular surface that partially or completely obscures the object being observed. Also called a veiling reflection.

Refraction

The bending of light as it passes between mediums of different densities

Reflector

The part of a luminaire that partially encloses the light source and redirects, through reflectance, some light emitted from it.

Reflector Lamp

A lamp in which a portion of the outer glass envelope is coated with a reflecting material to direct the light, also referred to as an R-lamp; variations include the ER and BR. Generically, this may also include other shapes that have reflectors such as MR and PAR.

Refractor

A device used to redirect the light output from a source, primarily by bending the waves of light (refraction).

Relay

An electrically operated switch.

Retina

A membrane lining the inner eye, opposite the pupil, which contains visual and non-visual photoreactive cells that transmit signals to various areas of the brain.

Retrofit

Installing new components into an existing luminiare

Return on Investment (ROI)

The benefit that results from the investment a resource. In construction this generally means the amount of time that will pass before the savings that result from an investment, as in a new energy efficient lighting system, equal the initial investment.

Rods

Retinal photoreceptors that provide vision at low light levels; rods do not provide color vision.

Roof Monitor

The raised section of a building's main roof; the sides of monitors usually contain clerestory windows or louvers to light and/or ventilate the area below.

Room Cavity Ratio (RCR)

A number computed by using the distance from the calculation plane to the luminaires that is used to quantify how light will interact with room surfaces; it is a factor used in illuminance calculations.

Schema

A mental framework or concept that helps us interpret and organize new information

Scotopic Vision

Vision that relies only on rods; scotopic vision occurs at illuminance levels below .1 lx.

Semi-Direct

A light distribution pattern or luminaire in which 60 percent or more of the light is directed downward and 40 percent or less of the light is directed upward.

Semi-Indirect

A light distribution pattern or luminaire in which 40 percent or less of the light is directed downward and 60 percent or more of the light is directed upward.

Semi-Recessed

A luminaire that is only partly above the ceiling or within a wall.

Semi-Specular

Describes a reflective material that is partially mirror-like. With semi-specular materials light is reflected directionally, but with some amount of scatter.

Shielding Angle

The angle measured from the ceiling plane to the line of sight where the bare light source in a luminaire becomes visible.

Side Lighting

Daylight illumination provided by windows mounted below the ceiling plane.

Simple Payback

The amount of time required for the savings on a new lighting system to equal the costs of that system.

Skylight

A glazed opening on the roof of a building.

Socket

The part of a luminaire that provides an electrical connection for a replaceable lamp and also holds the lamp in the proper position within the luminaire.

Solar Heat Gain

Heat added to the interior of a building by direct sunlight entering through windows or skylights.

Solid-State Lighting (SSL)

Lighting that uses light emitting diodes (LED) or organic light emitting diodes (OLED).

Spacing Criterion

The maximum distance that luminaires may be spaced while still providing uniform illumination on the work plane; the luminaire height above the work plane multiplied by the spacing criterion equals the center-to-center luminaire spacing.

Specification

In architecture, a written document providing information about the requirements of building components necessary to convey design intent and the legally enforceable contract requirements.

Spectral Power Distribution (SPD)

A tabular or graphic representation of a light source's radiant power at each wavelength.

Specular

Describes a mirrored or polished surface. On specular surfaces the angle of incidence is equal to the angle of reflection.

Spot Relamping

Replacing lamps as they fail on a one-by-one basis.

Steradian

An area of a sphere that is equal to the square of the sphere's radius.

Supply Charges

An item on a utility bill representing the costs of electricity a customer has consumes, typically in kilowatt hours.

Suprachiasmatic Nucleus (SCN)

The central pacemaker of the circadian timing system which regulates most circadian rhythms in the body.

Surface Mount

Describes any luminaire installed on the surface of the ceiling, wall, floor, or piece of cabinetry or furniture.

Task Lighting

Illumination provided to a specific area, such as a desktop, for the tasks performed there.

Task Tuning

The practice of adjusting the light output or spectral power distribution of the luminaires in a space to achieve and maintain the target illuminance or spectrum.

Thermal Factor

A factor used in lighting calculations that compensates for the change in light output of a discharge lamp due to a change in bulb wall temperature; it is applied when the lamp-ballast combination under consideration is different from that used in the photometric tests.

Top Lighting

Daylight illumination provided by skylights or roof monitors.

Transformer

A device that increases or decreases electrical pressure; in lighting this typically refers to a device that converts 120V to 12V.

Translucent

Describes a material that diffuses light passing through it so that objects cannot be seen clearly through it.

Transmission

The passage of light through a material.

Transmittance

The percentage of light that passes through a translucent or diffusing material.

Transparent

Describes a material with no diffusion of the light passing through it so that objects are clearly seen through it.

Tunable Lighting

Describes a luminaire or lighting system where the SPD can be adjusted, usually by an external control system.

Tunable White

A type of tunable lighting where the CCT can be adjusted.

Tungsten Halogen Lamp

See Halogen Lamp

Ultraviolet (UV)

Radiant energy with wavelengths between 10 and 400 nm.

Vacancy Sensor

A control device with which lights must be turned on manually, but that automatically turns lights off after the space becomes unoccupied; it may be ultrasonic, microphonic, and/or infrared technologies.

Veiling Reflection

See Reflected Glare

Visual Acuity

The ability to see fine or small details.

Visual Task

A vision related activity, such as reading a book, and the details and objects that must be seen for the performance of that activity

Volt

The unit of measurement for electrical potential; it defines the "force" or "pressure" of electricity.

Wall Box

Describes a wall mounted switch or dimmer

Wall Grazer

A luminaire set at the top of a wall that directs light down the face of the wall to illuminate it.

Wall Washer

A luminaire designed to illuminate a wall. They are usually used in groups and installed +/–3Êź from the wall.

Watt (W)

The unit for measuring electrical power; it defines the rate of energy consumption by an electrical device when it is in operation.

Wavelength

The distance from crest to crest, or trough to trough, of a wave. In light, wavelength is measured in nanometers (nm) or 0.000000001 meters.

Work Plane

The level at which work is done and at which illuminance is specified and measured; for office applications, this is typically a horizontal plane 30 inches (76 cm) above the floor (desk height).

Zone

See Control Zone

Index

A

accent 3, 85, 89, 90, 125, 126, 235
adaptation 21, 28, 29, 235
adjustable 89, 90
aesthetic design 15, 16
A-lamp 64, 67, 68, 209
alternating current 81, 142, 147
ambient luminescence 11, 12
angle of incidence 23, 124, 125, 235, 253
angle of reflection 253
aperture 86, 87, 219, 235, 236
AR-lamp 59
astronomical time clock 160, 236
audit 60, 195, 206, 225, 226, 229

B

baffle 88, 94, 214, 236, 246
ballast factor 132, 134, 236
ballast 69, 72, 134, 206, 222, 236, 242
beam angle 66, 67, 101, 236, 248
benchmark 192, 209, 229
binning 43, 81, 236
blackbody locus 35, 36, 37, 237, 241
blackbody 36, 38, 39, 51, 237
blade 88, 236
B-lamp 64
blue pump LED 77
BMS 158, 163
BREEM 187

C

CA-lamp 64
CCT 39, 42, 157, 161, 216, 239, 254
ceramic metal halide 73, 130
chroma shift 48, 49, 50, 51
chroma 34, 48, 49, 50, 53, 237
chromaticity diagram 33, 34, 35, 36, 37, 249
C-lamp 64
clerestory 237, 251
COB 80, 84
coefficient of utilization 127, 128, 129, 134, 237
Color rendering 5, 33, 45, 50, 182, 189, 227, 238
color temperature 5, 38, 39, 40, 238, 239
color vector 51
commissioning 10, 150, 153, 166, 238
compact fluorescent 61, 71, 209, 216, 238
conceptual design 16, 18
cones 30, 175, 239, 247
connected load 143, 164, 230
contrast ratio 8, 182
contrast 21, 29, 109, 239
control narrative 10, 165
control operations schedule 166
control schedule 141, 165, 166
control station 153, 154, 165, 239

control zone 108, 141, 152, 239
correlated color temperature 33, 36, 39, 40, 42, 238, 239
cost of ownership 6, 202
cove light 92, 239
CRI 33, 45, 46, 50, 182, 238
CRREM 192, 199
CT 38, 40

D

DALI 9, 82, 157, 240
daylight harvesting 104, 152, 161, 164
daylighting 103, 104, 106, 110, 111, 240
declare 183, 194, 195, 211, 221
delivery charges 208, 240
design intent 53, 55, 214, 252
designlights consortium 196, 200
DGNB 198, 199
diffuse 25, 88, 98, 99, 240, 247, 254
diffuser 25, 88, 97, 109, 240
diffusion 21, 95, 254
dimmer 143, 145, 154, 248, 255
direct current 78, 81, 142, 155
direct 11, 51, 63, 64, 98, 141, 236, 239, 244
downlight 16, 61, 74, 86, 88, 89, 123, 229
driver 29, 81, 148, 155, 207, 221
Duv 33, 40, 41, 241

E

efficacy 60, 61, 242
electromagnetic spectrum 21, 22, 171, 246
elements of quality lighting 5
ELV dimming 141, 148
energy legislation 208, 209
energy star 192, 197, 226
environmental product declaration 187, 194, 242

F

fenestration 104
fidelity 46, 48, 50, 52, 54, 238
field angle 66, 67, 101, 242
Fitwel 198, 199
flange 86, 87, 242
fluorescent 39, 60, 61, 68, 70, 71, 130, 216, 243
focal glow 11, 12, 13
footcandle 117, 243, 244
forward phase dimming 146, 147

G

G-lamp 64
glazing 104, 108, 110, 111, 113
graphic 14, 51, 115, 195, 236
green globes 190, 211
green star 198, 199
group relamping 207, 213, 217, 218, 219, 243

H

halogen 61, 62, 63, 130, 243, 248
HID 72, 74, 75, 216, 225, 236, 243
high-intensity discharge 72
high-pressure sodium 73
housing 83, 87, 193, 250
hue shift 48, 50, 53
hue 34, 49, 50, 53, 238
Hunt effect 27

I

illuminance 16, 27, 103, 110, 116, 126, 239, 243, 249

incandescent 2, 34, 39, 40, 61, 62, 130, 238, 243
indirect 7, 98, 99, 176
induction 75
Initial cost 202, 204, 206

L

L70 82, 131, 190, 219, 245
lamp failure 213, 216
lamp lumen depreciation 75, 130, 161, 219, 246, 247
lamp replacement 206, 207, 215
lamp shapes 60, 63, 64, 67, 70, 74
layers of light 10, 11
LED 17, 39, 40, 42, 59, 77, 95, 176, 217, 219, 241, 245
LEED 2, 8, 181, 188, 211
lens 25, 88, 90, 93, 248
level 1 audit 227, 228, 229
level 2 audit 227, 230, 231
level 3 audit 227, 230
life cycle assessment 194
life cycle cost analysis 201, 205, 206, 230
Light emitting diode 77, 80, 246, 248, 252
light loss factors 103, 113, 130, 132, 133, 134, 213, 246, 247
light shelf 108, 109
lighting application 2, 3, 13
lighting design process 3, 7, 13, 179
Lighting design 1, 2, 4, 7, 10, 13, 15, 18, 112, 191
linear fluorescent 61, 70, 130
LM-80 82, 83, 138
louver 88, 179, 214, 239, 246, 251
lumen method 126, 133
lumen power density 164
Lumen 41, 80, 99, 125, 126, 237, 243, 244, 246
Luminaire cleaning 206
luminaire dirt depreciation 131, 132, 247
luminaire maintenance 213
Luminaire 3, 9, 85, 97, 99, 161, 180, 206, 220, 247
luminaire-level lighting controls 162, 169
luminance 16, 104, 116, 126, 175, 243, 244
lux 30, 82, 117, 173, 175, 244

M

MacAdam ellipse 33, 43, 44, 45, 81
maintained illuminance 113, 126, 130, 132
matte reflection 21
mesopic vision 247
metal halide 39, 73, 130, 243, 247, 248
microphonic sensor 141
microwave sensor 141
MR-lamp 66

N

n-layer 78
nanometer 22, 246, 255
narrow band emitter 78, 248

O

occupancy sensor 144, 145, 150, 164, 177, 230, 248
OLED 77, 80, 248, 252
operating cost 83, 167, 201, 202, 210, 235
organic light emitting diode 80, 248, 252

P
PAR-lamp 59
passive infrared sensor 141
pc-LED 77, 78, 79
pendant 14, 28, 86, 97, 129, 176, 248
phosphor 68, 79, 218, 221, 222, 249
photocell 152, 158, 240, 248
photopic vision 21, 30, 249
PIR 151, 159
Planckian locus 36, 42
play of brilliants 11, 13
p-layer 78
PoE 157, 158, 249
point method 120, 121, 124, 125, 126
pragmatic design 14
preference 48, 54
preset 18, 144, 152, 154, 182, 238, 249
priority level 53, 55, 56
product ingredient list 187

Q
qualified products list 202, 210, 249

R
Ra 16, 48, 134, 159, 180, 249
rate of return 205
Rcs 52, 53
RDM 141, 156, 157
recessed 7, 88, 89, 96, 241
redesign 226, 230
reflectance 29, 50, 107, 116, 126, 250
reflection 21, 23, 24, 151, 179, 210, 245
reflector factor 133
reflector 23, 86, 116, 239
refraction 25, 250
relay 141, 145, 146, 165, 166, 250
retrofit 60, 69, 71, 137, 192, 201, 221, 225, 227
return on investment 205, 232
reverse phase dimming 72, 146
Rf 48, 52, 53, 189
Rg 52, 189
Richard Kelly 10
R-lamp 66, 250
rods 27, 30, 175, 247, 251
roof monitor 109, 110, 240, 254
room cavity ratio 128, 129
room surface dirt depreciation 132, 247

S
scene 152, 158, 167, 197
scotopic vision 30, 251
SCR 141, 147
semi-direct 98, 251
semi-indirect 98, 99, 251
semi-recessed 97, 251
semi-specular reflection 23, 24
side lighting 107, 108, 252
simple payback 201, 203, 204, 252
skylight 7, 103, 106, 110, 112, 161, 164, 240, 252
solar heat gain 105, 107, 252
solid state lighting 42, 78, 196, 252
spacing criterion 135, 252
SPD 33, 34, 38, 45, 49, 79, 253, 254
spectral power distribution 33, 161, 173, 237, 253
specular reflection 23, 24
spot relamping 207, 213, 217
supply charges 207, 208
surface mount 88, 89, 97
surface-mounted device 77
sustainability 8, 158, 181, 187, 190, 198, 219

switch 61, 141, 143, 145, 147, 150, 155, 156, 164, 214, 228, 229, 250

T

take back programs 84, 187, 193
task tuning 161, 162, 253
thermal factor 133, 134, 253
time scheduling 160, 164
T-lamp 64
TM-21 82, 83, 138
TM-30 27, 33, 48, 50, 52, 53, 56, 175, 230
top lighting 103, 107, 109, 254
track head 93, 94, 97, 214
track light 74, 88, 93, 122, 159
track 74, 93, 94, 192
TRIAC 147

U

UL 2809 194, 195
ultrasonic sensor 141
utility bills 207, 208
utility rebates 201, 204, 206, 209

V

vacancy sensor 9, 149, 254
visible spectrum 22, 105, 173, 248
visible transmission 113
visual performance 27, 120, 241, 243
vividness 27, 34, 48, 53, 54
voltage factor 134

W

wall grazer 91, 98, 255
wall washer 86, 90, 91, 94
wallbox 145, 150
wavelength 22, 34, 38, 39, 41, 79, 242, 253, 254, 255
white tuning 161
Wireless controls 158

Z

zone 78, 100, 111, 141, 152, 153, 164, 166, 182, 239, 255

About the Author

Jason Livingston, LC, IES, IALD is the principal of Studio T+L, a leading lighting design firm based in New York City. In addition to his design work, Mr. Livingston is on the faculty of New York School of Interior Design and New York University, and is the co-chair of the IES Color Committee. He is also the author of Designing with Light: The Art Science and Practice of Architectural Lighting Design. Mr. Livingston is a frequent speaker on lighting design and color related topics, especially color rendering and ANSI/IES TM-30.